Research for Development

Series Editors

Emilio Bartezzaghi, Milan, Italy

Giampio Bracchi, Milan, Italy

Adalberto Del Bo, Politecnico di Milano, Milan, Italy

Ferran Sagarra Trias, Department of Urbanism and Regional Planning, Universitat Politècnica de Catalunya, Barcelona, Barcelona, Spain

Francesco Stellacci, Supramolecular NanoMaterials and Interfaces Laboratory (SuNMiL), Institute of Materials, Ecole Polytechnique Fédérale de Lausanne (EPFL), Lausanne, Vaud, Switzerland

Enrico Zio, Politecnico di Milano, Milan, Italy
Ecole Centrale Paris, Paris, France

The series Research for Development serves as a vehicle for the presentation and dissemination of complex research and multidisciplinary projects. The published work is dedicated to fostering a high degree of innovation and to the sophisticated demonstration of new techniques or methods.

The aim of the Research for Development series is to promote well-balanced sustainable growth. This might take the form of measurable social and economic outcomes, in addition to environmental benefits, or improved efficiency in the use of resources; it might also involve an original mix of intervention schemes.

Research for Development focuses on the following topics and disciplines:

Urban regeneration and infrastructure, Info-mobility, transport, and logistics, Environment and the land, Cultural heritage and landscape, Energy, Innovation in processes and technologies, Applications of chemistry, materials, and nanotechnologies, Material science and biotechnology solutions, Physics results and related applications and aerospace, Ongoing training and continuing education.

Fondazione Politecnico di Milano collaborates as a special co-partner in this series by suggesting themes and evaluating proposals for new volumes. Research for Development addresses researchers, advanced graduate students, and policy and decision-makers around the world in government, industry, and civil society.

THE SERIES IS INDEXED IN SCOPUS

More information about this series at https://link.springer.com/bookseries/13084

Laura Montedoro · Alice Buoli ·
Alessandro Frigerio
Editors

Territorial Development and Water-Energy-Food Nexus in the Global South

A Study for the Maputo Province, Mozambique

Editors
Laura Montedoro
Architecture and Urban Studies (DAStU)
Politecnico di Milano
Milan, Italy

Alice Buoli
Architecture and Urban Studies (DAStU)
Politecnico di Milano
Milan, Italy

Alessandro Frigerio
Architecture and Urban Studies (DAStU)
Politecnico di Milano
Milan, Italy

ISSN 2198-7300 ISSN 2198-7319 (electronic)
Research for Development
ISBN 978-3-030-96540-2 ISBN 978-3-030-96538-9 (eBook)
https://doi.org/10.1007/978-3-030-96538-9

This Springer imprint is published by the registered company Springer Nature Switzerland AG
The registered company address is: Gewerbestrasse 11, 6330 Cham, Switzerland

About This Book

The safeguarding of primary resources such as land, water, air, food, energy, and the right to housing challenges societies to regulate the ratio between ecological footprint and development, looking for more sustainable, resilient, and inclusive processes of urbanization.

These challenges require innovative design and governance tools capable of integrating aspects of variability and/or informality that represent a paradigm shift with respect to the consolidated European planning tradition.

How to increase people's quality of life by keeping territorial development sustainable and durable?

To study complex and rapidly changing urban settings "such as in Sub-Saharan Africa, where the social, cultural, economic, environmental and physical contexts are very different from those in the post-industrial urbanized 'North' (Smith and Jenkins, 2015) a transdisciplinary approach is needed and it is discussed in the contributions of the volume, from different perspectives.

The volume sets the research framework and challenges through different and complementary cultural and disciplinary gazes applied to the specific case study of Maputo, Mozambique, supported by specific insights from a research by design project.

The volume originates from the research results of "Boa_Ma_Nhã, Maputo!" (Polisocial Award 2018), a transdisciplinary research project based at Politecnico di Milano and focused on the districts of Boane, Moamba, and Namaacha, located in the eastern part of the Maputo Province, Mozambique. The project has been a testing ground to integrate and blend different know-hows and backgrounds when dealing with contingent issues. It involved four departments from different fields (architecture and urban studies, environmental, electronics, civil and energy engineering) and has been developed in partnership with the Eduardo Mondlane University (Maputo, Mozambique) and the Italian Cooperation Agency (AICS) in Maputo. It also benefits from the support of other locally operating NGOs and institutions and is framed in a larger international cooperation initiative named PIMI (Programa de Investigação Multisectorial Integrada: Estudo para a Promoção do Desenvolvimento Territorial Integrado da Região de Boane, Moamba e Namaacha).

"Boa_Ma_Nhã, Maputo!" project questions the role and capability of existing and potential/alternative planning tools in determining sustainable and resilient territorial development, with special attention toward the vulnerabilities and the emerging challenges observed in the area. Among them, the most evident and urgent are the changing rural–urban socio-economic balance; the local effects of climate change; water competition; food insecurity, and access to energy. All these conditions make the peripheral districts of the fast-growing metropolitan area of Maputo a fragile territory in need of appropriate tools to strategically drive sustainable growth in a synergic, resilient, and inter-scalar perspective.

Research activities include desktop and fieldwork research, production of cartographies, data model processing, and scenarios definition, and are conducted by adopting multidisiplinary and interdisciplinary approaches, for the implementation of an integrated transdisciplinary assessment, a WEF-sensitive planning framework, and a local development project to test potential and effective implementation processes of the general strategy.

In addition, specific attention has been devoted to the co-production of knowledge by activating dialogs across disciplines, with the project partners, and different local and international actors.

The volume capitalizes on this research by design experience to contribute to the international discourse on transdisciplinary and transcultural urban and territorial studies regarding sub-Saharan African cities, highlighting inventive patterns in methodological, analytical, and design terms.

Contents

Presentation

Emanuela Colombo and Manuela Nebuloni

1 Mozambique, Africa and Polimi: Building Integrated Solutions

Over the last decades, universities have proven their ability to support development and equity through innovation generation. Their contribution became a qualifying element for the European Union's foreign policy, while in Italy, academia is recognized as an actor in the realm of cooperation by the law 125/2014: it is within this context that the personal initiatives of some academicians at Politecnico di Milano have been transformed into an institutional vocation, now unfolding according to the goals and recommendations of the 2030 Agenda of the United Nations and Agenda 2063 of the African Union, with a people-centered approach. The strategic direction in the area of cooperation and development at Polimi can be summarized in three areas: Higher Education and Capacity Building, Research for local Development, and Science Diplomacy.

Besides the traditional pillars of research and teaching, partnerships and networks enforce the dialog between universities and the outside world; by contributing to responding to needs identified outside of the academic realm, University research is enriched by an approach centered around multi-disciplinarity and mutual learning.

To further enforce the institutional engagement of Politecnico di Milano within the frame of Responsible Research—including cooperation actions—in 2013, our university launched the Polisocial Award initiative, a self-funded competition in

E. Colombo
Rector's Delegate to Cooperation and Development, Politecnico di Milano, Milan, Italy

UNESCO Chair in Energy for Sustainable Development, Department of Energy - Politecnico di Milano, Milan, Italy

M. Nebuloni (✉)
Polisocial Award Organising Committee, Public Engagement and Communication Unit, Politecnico di Milano, Milan, Italy

© The Author(s), under exclusive license to Springer Nature Switzerland AG 2022
L. Montedoro et al. (eds.), *Territorial Development and Water-Energy-Food Nexus in the Global South*, Research for Development,
https://doi.org/10.1007/978-3-030-96538-9_1

1

support of scientific research with a high social impact, made possible by the many donations through 5 per Mille IRPEF. Such initiative wishes to promote:

- Multidisciplinary research is able to create synergies to address complex issues of relevance to communities for human and socio-economic development.
- Innovation, through the placement of research activities in problematic contexts, with the aim of developing methods and knowledge of more general applicability.
- Dialog with the outside world and the co-generation of knowledge, through the creation of stable partnerships with institutions, companies, civil society, and international organizations, in which the role of the University is enhanced as an expert interlocutor and coordinator of research activities on issues of direct interest to multiple communities.

The Polisocial Award 2018 was dedicated to the theme "Sustainable Cities and Communities in Africa": four projects were funded, each proposing multidisciplinary research for enhancing local development.

"Boa_Ma_Nhã, Maputo!" was one of the winning projects and one of the two funded initiatives in Mozambique. The research was carried out by four departments (Department of Architecture and Urban Studies; Department of Electronics, Information and Bioengineering; Department of Energy; Department of Civil and Environmental Engineering) and a multi-actor partnership.

The peculiarity of "Boa_Ma_Nhã, Maputo!" consists of the multidisciplinary approach to the cooperation between architects and engineers. The researchers worked toward the definition of cross-cutting methodologies to guide the local decision-making and planning processes, through the construction of scenarios and recommendations on water, energy, food, and urban development at a territorial scale.

Politecnico di Milano's interest in Africa, in general, and Mozambique, in particular, is confirmed by the recent mapping of all Cooperation experiences activated by our university since 2011: Africa is at the center of the partnerships and action research at Polimi being the focus of more than 60% of the over 100 mapped initiatives. Additionally, projects involving African countries are three times those involving Asia and six times those involving Latin America. Twenty-five African countries have been the research ground of such initiatives, with the recurring presence of 10 nations: Mozambique, Tunisia, Egypt, Kenya, Ethiopia, Morocco, Senegal, Somalia, Tanzania, and Algeria.

Mozambique is a strategic partner for Italy, confirmed as such also for the period 2020–2022, according to the strategy of Cooperation and Development recently approved. In line with the national priorities, as in the case of "Boa_Ma_Nhã, Maputo!", the country has often been approached by Polimi's researchers with the support of local partners and under the lenses of different disciplines, with strong attention to its territorial autonomous development.

Among the peer projects self-funded by Polimi in Mozambique:

- Mo.N.G.U.E., an investigation into the condition of fragility and potential of a territory, with regards to the environment and education (Polisocial Award 2016).

- SAFARI NJEMA, research dedicated to informing the mobility policies through big data analysis, to promote bottom-up replicable solutions (Polisocial Award 2018).
- HANDS, addressing the promotion of a culture of healthiness through multidisciplinary experimentation of medium and long-term actions in informal urban settings (Polisocial Award 2020).

Initiatives activated thanks to the collaboration with AICS Maputo:

- AILs—African Innovation Leaders, which trained 21 young African professionals (2 from Mozambique) on issues related to the Next Production Revolution, entrusted with training a second-generation of innovators (2018–2019).
- Monitoring and assessing the impact of ILUMINA, a rural electrification program in the regions of Zambezia and Cabo Delgado (2019–2022).
- ICT4DEV, training for students, professors, and researchers of the Eduardo Mondlane University in the ICT sector (2021–2023).

EU-funded initiatives involving several African countries, including Mozambique, are part of the Horizon 2020 projects DAFNE, associated with the management of water resources in complex transnational basins and LEAP-RE, related to the promotion of renewable energy in Africa, boasting a partnership of over 80 institutions from 34 African and European nations.

The creation of partnerships to develop all these projects is grounded on the idea that Development needs to be an integrated process: primary resources and common goods such as land, water, air, food, energy, and the right to housing are interconnected and call for a more resilient urban development and a more sustainable transformation of communities in rural areas, engaging the African genius in the promotion of native innovation in both the manufacturing and the agriculture sectors and in community services like health and education.

Introduction. A Polytechnic Approach to Urban Africa. Methodological and Cultural Challenges of a Transdisciplinary Research Cooperation

Laura Montedoro, Alessandro Frigerio, and Alice Buoli

Abstract This introductory chapter provides and discusses the main cultural and methodological frameworks in which the contributions of the volume are embedded. Drawn on current debates on academic partnerships between international educational and research institutions, the essay explores the increasing importance of pluralism, internationalization, and new policies of social responsibility in performing transdisciplinary research in the Global South. The text also discusses the specificities and originality of the approaches adopted in the past decades by the book authors' institutions—the Politecnico di Milano and the Eduardo Mondlane University (UEM)—embedded into the mutual interplay between space and society, multidisciplinary and practice-based research, and the role of socially and environmentally responsible transformations, in dialogue with other academic approaches to international cooperation. The paper further presents the ongoing academic joint initiatives between Polimi and UEM, introducing two major research and educational programmes that are the main source of evidence for this volume's contributions and discussion within and beyond the international academic networks involved. In addition, an overview of the book's structure and main contents is provided.

L. Montedoro · A. Frigerio · A. Buoli (✉)
Department of Architecture and Urban Studies, Politecnico di Milano, Milan, Italy
e-mail: alice.buoli@polimi.it

L. Montedoro
e-mail: laura.montedoro@polimi.it

A. Frigerio
e-mail: alessandro.frigerio@polimi.it

© The Author(s), under exclusive license to Springer Nature Switzerland AG 2022
L. Montedoro et al. (eds.), *Territorial Development and Water-Energy-Food Nexus in the Global South*, Research for Development,
https://doi.org/10.1007/978-3-030-96538-9_2

5

1 Situating International Educational and Research Cooperation in the Global South

As suggested by Petrillo and Bellaviti (2018), international cooperation initiatives among universities and research centres working on territorial and urban development and planning have assumed a more and more prominent role in view of the challenges posed and the opportunities offered by the UN "New Urban Agenda" (UN-Habitat 2015; UN 2016). Indeed, during the past decades, academic partnerships have been shaped by the increasing importance of pluralism, internationalization, and new policies of social responsibility

> which aim to make room for relating, exchange and sharing of knowledge and skills among a wide variety of subjects (students, lecturers, researchers, institutions, the private world, NGOs, civil society) and an increasingly broad range of countries, cities, territories, building up networks of knowledge and training that move beyond geopolitical and cultural frontiers (Petrillo and Bellaviti 2018: xiv).

At the same time, a major challenge today appears to be overcoming the paradox of a widespread scientific literature on African urbanism produced in the Global North, with low production (but often just a limited visibility) from the Global South, where in many cases there's a lack of urban planning and management professionals accordingly trained to respond to the rapidly evolving urban complexity. Paul Jenkins stresses the North-centric prevalence of research activities in his "Urbanization, Urbanism and Urbanity in an African City" (Jenkins 2013) and Vanessa Watson and Babatunde Agbola, wondering "Who will plan Africa's cities?" (Watson and Agbola 2013) stress how "the urban and rural planning curricula of many planning schools are as outdated as planning legislation" and, moreover, some African countries even don't have a planning or architecture school.

However, to limit the dominance of "unsuitable archetypes" (Watson and Agbola 2013), related to colonial legacies or post-colonial cultural dependencies, the potential role of academic institutions as inclusive places for multi-directional North-South exchange, cooperation, and co-production of knowledge and awareness is crucial.

In the last decade, in fact, the interest of the most important technical schools at the global scale towards African territorial development and planning issues has been raised. This happened for a combination of reasons interweaving research interests with political-economic interests linked with multilateral or bilateral cooperation targets, but it hasn't always implied a fair involvement and exchange with local academic institutions. Working together, instead, should have the crucial aim of setting alternative balanced cooperation frameworks, able to take advantage of the available resources with the shared objective of exchanging knowledge and building on it to produce common contemporary-aware, transculturally responsive, and locally sensitive theories and methodologies. This is anything but easy in practice, but it is of particular importance when considering design education and polytechnic disciplines (including urban planning, engineering, and architecture) not only as regards scientific and institutional knowledge exchange, but also for the implementation of local development projects and plans, as well as hands-on educational programmes facing the challenges of contemporary development trends and conditions in different geographical contexts, characterized by a tangled combination of global and local dynamics.

In this framework and considering a long- and well-established relationship between Mozambican and Italian (academic) institutions since the mid-70s, during the last decade, the Eduardo Mondlane University - UEM[1] (Maputo, Mozambique) and Politecnico di Milano[2] (Milan, Italy) have been involved in higher education cooperation programmes, research initiatives, pedagogical exchanges, capacity building and training on relevant technical fields such as mobility and water management, food security, energy production, urban planning, along with spatial planning and architectural research and design.

The two institutions share a common and well-established attention to international agendas and a special care in localizing issues, as well as in spatializing them in physical terms with an experimental research-by-design attitude relying on the powerful neutral role of academia (out of political or entrepreneurial interest). The specific approach of the two academic institutions has a key role in setting the value of their cooperation.

Politecnico di Milano (Polimi) has a solid polytechnic vocation to recognize and strategically exploit territorial values by combining synergic skills in architecture, urban design, planning, urban studies, and civil, environmental, management, energetic engineering, among many. Research on territorial development and strategic planning, including projects, plans, studies, policies, and processes has a long tradition and blends different perspectives in terms of scale, methodology, and tools, with the general character of high respect for local values in their tangible and intangible assets. As regards the Department of Architecture and Urban Studies (DAStU), their main expertise can be recognized in the careful reading of territories, according to both cultural and technical lenses, with the aim of conceptualizing the interdependence of social and physical relational patterns in terms of robust structure of urban settlements to be protected and strengthened. This effort of conceptualization is key in strategically supporting democratic and inclusive policymaking and decision-making processes as agents of sustainable transformation and development.

The Eduardo Mondlane University has a strong rooting in the territory of Maputo metropolitan region, being involved in the elaboration and technical assistance for the design of planning tools and projects for the local governments of the area and for the city of Maputo itself. In particular, the Faculty of Architecture and Planning (FAPF) hosts the Centre for Development of Habitat Studies, a research and services institution created in 1992, whose objectives are, among others, contributing to the improvement of habitat conditions in Mozambican cities. In addition, in the context of collaborations with international research institutions and centres (i.e., the University

---

[1] Funded in 1962, the Eduardo Mondlane University (Portuguese: Universidade Eduardo Mondlane; UEM) is the oldest and largest university in Mozambique. Today it counts more than 30 faculties and research centres, and has about 40,000 students enrolled—www.uem.mz.

[2] Politecnico di Milano (Polimi) established in 1863 is ranked among the top 20 universities in the world QS World University Rankings by Subject 2020 in the category "Architecture" (Ranking #7) and Design (Ranking #6), "Engineering & Technology" (Ranking #20)—www.polimi.it.

College London—UCL or the Danish Research Council) the FAPF has been involved in the development and testing of climate change adaptation/mitigation tools and methods for urban development (Broto et al. 2015).

The most relevant and recent cooperation initiatives between Polimi and FAPF-UEM have been carried out in the context of the Politecnico social responsibility programme (the Polisocial Award[3]) and the memorandum of understanding between the Italian Cooperation Agency (AICS), the UEM, and several Italian universities in the field of academic restructuring, scientific research, and technological innovation. In particular, the Polisocial programme has been designed and carried out since 2011 with the aim to introduce

> a new way of building and applying knowledge and academic excellence, by fostering and supporting new multidisciplinary projectuality, aware to human and social development, and by widening training, as well as exchange and research opportunities, offered to students, young researchers, university staff and its network (Colombo et al. 2016: 5).

For this purpose, the AICS-UEM-Polimi trilateral partnership represents a promising setting in terms of long-term cooperation and funding scheme for applied research as well as educational innovative formats. Despite some inevitable criticalities due to bureaucratic difficulties in translating formal agendas and agreements into actual operative programmes, the partnership—whose implementation is underway and will be further enforced in the upcoming years—has so far allowed to carry out a number of short-term academic research cooperation initiatives and a long-term research-based educational programme, involving different units and departments from both universities, also in cooperation with other local and internal partners.

Starting from such premises, this book capitalizes on the research and educational cooperation "ecologies" between Italy and Mozambique and proposes a research-by-design and an interdisciplinary collaborative methodology to contribute to the international discourses on transcultural urban and territorial studies regarding sub-Saharan African cities, highlighting inventive patterns in methodological, analytical, and operative terms. To this aim, the volume sets the research framework and challenges through different and complementary cultural and disciplinary gazes applied to the specific case study of Maputo (Mozambique) supported by insights and tools from a research-by-design perspective. Contributions come from different departments in Politecnico di Milano and in Eduardo Mondlane University, with insights from various disciplines such as architecture and urban studies; geography; social sciences; civil and environmental engineering; electronics, information, and bioengineering; energy engineering.

In the following paragraphs, we introduce the main conceptual and theoretical framework on which the book's contributions are based and more specifically the WEF (Water-Energy-Food) nexus as the main conceptual and interpretative lens and source of scholarship for the interdisciplinary research projects and trajectories discussed in the volume. We, then, provide some general introductory information about the context of the Maputo Province, through two main Polimi-UEM research

---

[3] www.polisocial.polimi.it

and educational and research initiatives that are among the major sources of evidence for this book.

Finally, we provide some details about the structure of the volume and a synthetic overview of the contributions presented.

2 Cultural and Theoretical Framework and Know-How

The multidimensional and cross-disciplinary approach adopted by the research experiences presented in this book is embedded into sustainable development debates inside the field of urban planning, especially in African urban studies, as well into civil, environmental, and energy engineering.

The first mainframe of reference is related to UN-Habitat rural-urban linkages and partnerships in the Global South, following the UN Sustainable Development Goals and the New Urban Agenda launched on the occasion of the Habitat III conference in Quito (2016). According to UN-Habitat (2017).

> the concept of Urban-Rural Linkages contains the idea of complementary functions and flows between rural and urban territories of various sizes, such as metropolitan regions, small- and medium-sized cities and market towns as well as sparsely populated areas with the smallest scale of human settlements.

Such a framework is particularly appropriate when studying fast-growing metropolitan regions in sub-Saharan African countries: among them the case of Maputo represents a meaningful example in light of the current socio-economic interactions occurring between the attractive and vibrant capital of Mozambique and the small and medium-sized (r)urban centres gravitating around the city, witnessing very diverse urbanization and demographic patterns and trends, as well as manifold challenges related to the access to food, basic services and resources for the local populations.

The second source of reference (both in cultural and political terms) is the broader discussion about the challenges of urban planning and governance in Sub-Saharan African cities. Different scholars have tried to systematize scientific contributions on the characters of Sub-Saharan extreme urbanization (Mbembe and Nuttall 2008; Myers 2011; Pieterse and Simone 2013), with the recurring of keywords such as postcolonialism, cosmopolitism, informality, climate change, political and social fragility, governance weakness, invisibility, segregation, and gentrification. Their effort witnesses (and in most cases orients) research trajectories that are mainly focussed on a socio-cultural perspective investigated through a phenomenological lens. A relevant role, in this perspective, is currently played by the African Centre for Cities (ACC) at the University of Cape Town (UCT), which has become a hub for African researchers and a champion of an African perspective in producing knowledge in the field of urban studies (Harrison 2006).

This background is crucial in shaping an attitude of respectful sensitivity towards local contexts as well as an honest openness in dealing with unprecedented challenges. However, it's relevant to highlight limited attention to the urban and territorial physical dimensions. Differently, this attention emerges in the work of Behrens and Watson (1996), Jenkins (2013), Todeschini (2014), but it especially found fertile ground in the experimental attitude of local and international universities cooperating through workshops and teaching activities in the field—as witnessed, for instance, by activities by Columbia University (Kurtak and Daher 2011; Blaustein et al 2012) and Penn University (Gouverneur 2014)—or through specific initiatives such as the International Design Collaboration for Kenya promoted by UN-Habitat in 2016 (UN-Habitat 2016).

Finally, and most importantly, a major source of scholarship, connected to the previous ones, is related to the broad field of the so-called Water-Energy-Food (WEF) nexus. The nexus approach, embraced by FAO (2014) and many other important international organizations, has undergone different conceptualizations, according to scope, objectives, and understanding of drivers. However, the conceptual framework that systematically links natural environment and human activities through trade-offs and synergies is a crucial lens to investigate the complexity of territories and to set more integrated and cost-effective decision-making and planning processes by challenging existing borders, policies, and procedures at the various scales.

Scholars also underline how "whilst the WEF nexus scholarship has expanded since 2013, there is also evidence of growth in the conceptual, intellectual and social structures of the WEF nexus in the African continent" (Botai et al. 2021). At the same time, most of the existing literature about WEF Nexus and sustainable development in the Global South (Hassan Tolba et al. 2018; Botai et al. 2021; Purwanto et al. 2021; Wahl et al. 2021), however transdisciplinary, has a preferential methodological and conceptual perspective focussed on one of the elements of the nexus and/or limited to a very technical and non-spatialized approach. The aim of this volume, also building on previous research products on these topics by Politecnico di Milano (Colucci et al. 2017), is to investigate the nexus by starting from a spatial perspective. Therefore, it focuses on the effects of the nexus technical assessments on the physical space we inhabit with an integrated territorial/urban gaze. Moreover, the volume presents the challenges of uncovering and implementing the WEF Nexus in a context (Maputo, Mozambique, as representative of a large part of sub-Saharan African urban areas) that is still scarcely sensitive to these issues.

Looking at the Maputo Province and the critical issues related to the urban-rural dynamics pushed by climate change and socio-economic drivers, the transdisciplinary perspective given by the framework of the WEF nexus together with the main debates on urbanization trends and territorial development in African metropolitan regions, have profoundly shaped the interpretative and operative lenses adopted by the research projects presented in this volume.

3 Learning by Practice: Designing an Integrated Territorial Vision for the Maputo Metropolitan Area

The main territorial framework to which this volume is devoted is the metropolitan region of Maputo, and, in particular, the districts and municipalities of Boane, Moamba, and Namaacha. This area is in the south-eastern section of Mozambique and of the Maputo Province, bordering two other nations—South Africa and eSwatini—and it is crossed by the "Maputo Corridor" (Maputo—Johannesburg—Durban).

More than 3,000,000 inhabitants live in this urban agglomeration, representing more than 13% of the Mozambican population and over 40% of the urban population of the country (see Fig. 1). However, the lack of information regarding existing cross-scalar patterns that have been shaping this territory in the past decades makes Maputo an "unknown metropolis", fragmented in terms of administrative boundaries and territorial governance and shaped by a complex tangle of informal or unmapped flows and systems.

Among the main challenges that affect this territory, some are more evident and urgent, such as the changing rural–urban balance and social transitions due to demographic growth, migrations and progressive (and mostly unplanned) urbanization patterns, the local effects of climate change, deforestation and the following changes in the landscapes, food and water insecurity, land grabbing, socio-economic and political instability. All these conditions make the peripheral districts of the fast-growing Maputo metropolitan area a vulnerable territory in need to be framed in a synergic inter-scalar vision for a sustainable and integrated territorial development.

Along with these emerging issues, the scarcity and inconsistency of the available statistical data, the lack of published cartographic documentation and easily accessible digital databases and cartographies, and the scarcity of investigations of economic related transformations pose a series of challenges also in terms of research methodologies, planning and governance tools to be developed in support of local actors to face this crucial task while coping with present urgent issues.

The contributions presented in this volume—based at Polimi and FAPF-UEM—have embraced these challenges by proposing a multidisciplinary and multi-level planning approach to tackle the development of the growing peri-urban environment of Maputo in an integrated way, overcoming the traditional sectorial approach, and considering the interdependencies between issues such as migrations and demographic trends, unplanned urban growth, food and water security, climate change and natural hazards, local economic patterns (formal and informal), land tenure and cultural diversity, mobility, and infrastructure.

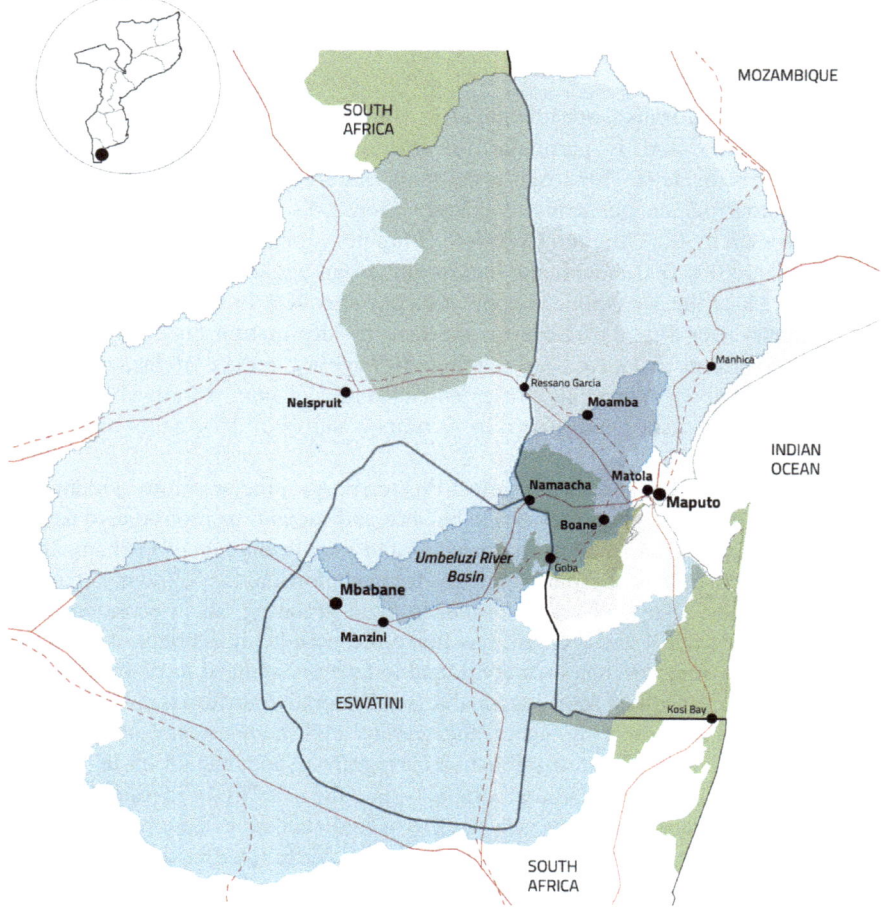

Fig. 1 Transnational research framework: administrative boundaries, main connections, protected areas, and river. basins. *Source* Elaboration by "Boa_Ma_Nhã, Maputo!" team 2019

3.1 From PIMI to "Boa_Ma_Nhã, Maputo!" and Back

As Carlos Trindade, João Tique, Domingos Macucule explain in their contribution, PIMI[4] is an ongoing research and educational programme based at the FAPF-UEM. The programme is part of the Intergovernmental Agreement between Italy and

---

[4] Acronym of *Programa de Investigação Multissectorial Integrada: Estudo para a Promoção do Desenvolvimento Territorial Integrado da Região de Boane, Moamba e Namaacha*, in English "Integrated Multisectoral Research Programme: Study for the Promotion of the Integrated Territorial Development of the Boane, Moamba and Namaacha Region".

Mozambique (2011) to support UEM in terms of academic and technological inno-vation and scientific research[5]. In the context of such programme, the FAPF, together with other units from UEM, is carrying out the programme within a timeframe of three years starting from 2020. The study aims to gain a thorough and in-depth understanding of the territory of the Boane, Moamba, and Namaacha region in its different dimensions, with a focus on the creation of an updated territorial database and the development of territorial planning and management activities. Thus, among the main aims of PIMI there is an extensive analysis, from a critical perspective, of the trends of transformation of the city of Maputo and the Boane, Moamba, and Namaacha region to further propose specialized tools for the protection and main-tenance of the cultural, environmental heritage in a sustainable perspective, as well as for intervening on the built and natural heritage in consolidated urban centres and fragile areas.

The research programme is articulated into five research lines: "Economy and Territorial Development"; "Territorial Planning"; "Socio-demographic and socio-territorial dynamics"; "Governance and Public Policies"; "Environment and Sustain-ability". From a general methodological perspective, the programme combines both research and educational activities in an integrated way. This means that the five different research lines have been assigned to five leading researchers that coordinate teams composed of other colleagues, Ph.D. students, and Master students from the faculty. Research activities—such as data collection and other research operations—overlap and integrate training activities (intensive workshops, seminars, lectures) throughout the programme.

Drawn on the territorial and conceptual framing defined by PIMI, in 2019, the research programme "Boa_Ma_Nhã, Maputo! A study for the integrated develop-ment of the region Boane, Moamba, Namaacha (Mozambique)" based at and funded by Politecnico di Milano, was established running throughout 2019 and 2020. The project has been carried out in partnership with the FAPF-UEM, the Italian Cooper-ation Agency (AICS) in Maputo, and the NGO Progetto Mondo MLAL. As already mentioned, the project is framed into international cooperation initiatives between Polimi, UEM, and AICS and has been designed to provide scientific support to PIMI with a specific interdisciplinary, cross-thematic, and multi-scalar approach, in order to fill in the knowledge gaps and co-produce new knowledge in support of future research and planning activities for the programme.

Indeed, particular attention has been devoted to the WEF Nexus, considering the potential evolution of the agriculture sector, backbone economy of the Maputo metropolitan region, and the related food systems in their multiple environmental, economic, social, and cultural implications. The project also assumed the UN Sustain-able Development Goals (SDGs) (UN 2015) and the UN-Habitat framework for urban–rural linkages as a main cultural and policy-oriented framework and reference (UN-Habitat 2017).

---

[5] See for more details: https://www.uem.mz/index.php/noticias-recentes/1119-uem-em-seminario-de-apresentacao-de-resultados-de-investigacao-do-fiam. Accessed 13 December 2021.

The main aim of the project has been to test an integrated, replicable, multi-disciplinary methodological approach to produce scenarios and specific guidelines to support decision-makers dealing with the challenges of sustainable development in fragile contexts of the Global South. The project further aimed at verifying the methodological approach through a locally relevant pilot project, involving local actors, and investing in education and local rural entrepreneurship with the aim of producing measurable impacts.

The project involved four different departments from Politecnico di Milano (Architecture and Urban Studies—DAStU; Civil and Environmental Engineering—DICA; Electronics, Information, and Bioengineering—DEIB; and Energy—DENG) which are all represented in this volume, together with different departments and faculties from UEM.[6] Among the main outcomes of the project, there are an analytical and interpretative transdisciplinary study, integrating remotely collected data and empirical on-site observations, on the current conditions of the Maputo Province, and a planning-oriented document entitled "Territorial guidelines and scenarios articulating the WEF Nexus in the Greater Maputo Region" which aimed at suggesting a series of strategic scenarios and actions for the territory of study. These ones represent the main contribution of the project as an interpretative and imaginative framework for supporting further research trajectories, first of all, PIMI, and policy-making initiatives by local authorities and the urban governance stakeholders in the Boane, Moamba, and Namaacha area, as well as for the entire Maputo metropolitan region.

3.2 Working Together: Methodological Challenges and Insights

Despite PIMI and "Boa_Ma_Nhã, Maputo!" having been conceived and designed as autonomous research programmes, along with the territorial focus, the projects share some key methodological features and tools. As mentioned before, the access to recent and systematized data and documentation has been from the very beginning a critical challenge for both Polimi and UEM research teams, due to the difficulties in collecting updated and coherent information both in terms of level of geo-referenced accuracy and details, administrative definition and formal (planning) documentation. For this reason, both research teams proceeded in a tentative and heuristic manner, by combining different methods and tools for data collection and representation, scenario-building and urban planning experimentations, as well as in terms of interaction with local agents and institutions (See Fig. 2).

In this regard, a key research operation has been the spatialization and visualization of the main territorial conditions and processes occurring at different scales (demographic trends and urban growth, land cover and land uses, food production

---

[6] Along with FAPF, also the Faculty of Arts and Social Sciences, the Department of Geography and the Centre for Policy Analysis (CAP) are represented.

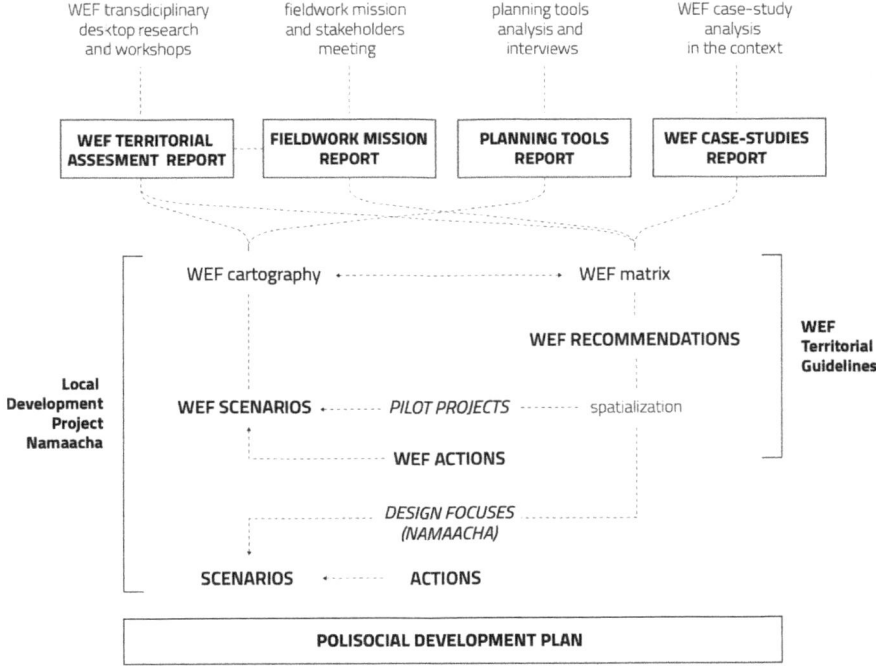

Fig. 2 Research operations: relations between methods, tools and outcomes. *Source* Elaboration by "Boa_Ma_Nhã, Maputo!" team, 2020

patterns, climate data, hydrographic systems, energy networks and resources, etc.) into synthetic cartographies and data-visualization diagrams. An in-depth description of the methodological approach and research results is provided in Part 1 and Part 2, devoted to the projects.

As mentioned earlier, both programmes adopted a research-by-design approach regarding the territorial guidelines and scenarios for the Maputo metropolitan region: the joint research efforts performed by the projects' partners have been oriented towards the provision of some WEF-sensitive and transdisciplinary policies and planning strategies to deal with the challenges and opportunities observed.

In this sense, for the purpose of both PIMI and "Boa_Ma_Nhã, Maputo!", a set of the research operations have been performed integrating educational activities inside the research process itself, providing training to MA and Ph.D. students, who were able to take on a proactive role and assume specific responsibilities.

The project teams also adopted Participatory Action Research methods, which emphasizes the co-generation of knowledge between the researchers, the projects partners, and other actors to implement and assess the result of the research process with the academic/scientific community of reference. On the occasion of both the fieldwork missions, a series of interviews with a number of local actors and experts have been organized. The interviews were mostly conducted in unstructured and

Fig. 3 Meeting at the "25 de Setembro" agriculture cooperative, Boane, August 2019. *Photo* by "Boa_Ma_Nhã, Maputo!" team, 2019

informal settings, allowing the respondents and the researchers to interact freely and add issues and questions during the conversations (See Fig. 3). This allowed the team to assess and re-orient the early findings and insights as regards the territory of the study.

The ongoing joint research and educational initiatives implemented by Polimi and UEM represent a meaningful example of academic cooperation in contributing to building knowledge and awareness about the challenges and potentialities conditioning urban development in a critical context, also providing an interpretative and imaginative framework for supporting further research trajectories and policymaking initiatives by local authorities and the urban governance stakeholders in the area.

Since the early implementation of PIMI, and later throughout "Boa_Ma_Nhã, Maputo!", the construction and maintenance of mutually proactive relations and partnerships between the two institutions has been key for the successful completion of the main research agenda of the two programmes. The interdisciplinary, inter-cultural, and multilingual nature of the projects has been, since the early moves of the programme, both a challenge and an opportunity for their development. The success of such "translation" operations—also in terms of communication outside the academic arena—is yet to be assessed: however, the effort to identify and design (cultural and disciplinary) "mediation tools" (such as data-visualization tools and interpretative matrices—Cfr. Part 2) can be seen as a positive output of the interactions among the projects' teams, especially in methodological terms.

The key role of academic international cooperation in reconceptualizing spatial planning for a more context-sensitive, equitable, and sustainable development needs this effort of setting open and inclusive common grounds.

4 How This Volume Is Structured

As mentioned, this volume is the outcome of the ongoing collaborations among Polimi and UEM and the long-term conversations with the colleagues that generously have contributed to building a fertile common ground for the co-production and sharing of knowledge, not only collaborating in the research activities conducted in Maputo, but also in the many occasions of encounter and discussion in the past three years that have involved a "community" of many different voices and expertise. In this sense, the book is a representative collection of the excellent scientific expertise which characterize the two institutions in different fields of knowledge: from social sciences to hydraulic, environmental, and energy engineering, to architecture and urban planning and design.

Benefitting from and exploiting such richness of viewpoints and expertise, the volume is organized into three main sections:

Part 1 (Chapters "Introduction. A Polytechnic Approach to Urban Africa. Methodological and Cultural Challenges of a Transdisciplinary Research Cooperation" to "The Demography of the Maputo Province") presents contributions by scholars from different disciplines based in Maputo and Mozambique introducing the main issues and challenges for this territory with context-based and long-term perspectives. These essays provide privileged entry points into research and innovation trajectories in urban history and planning, local material culture and habitat, societal challenges, and natural resources management, through the voices of prominent scholars based in Maputo and, in particular, at the Eduardo Mondlane University (UEM). This section includes essays by Paul Jenkins, Inês Macamo Raimundo, Domingos Macucule, João Tique, and Carlos Trindade.

Part 2 (Chapters "Integrated Multisectoral Research Programme (PIMI). Origins, Trajectories and Horizons" to "Energy-Food Challenges and Future Trends in Mozambique and in the Maputo Province") is devoted to presenting the main outcomes of the "Boa_Ma_Nhã, Maputo!" project, as an applied research-by-design case study, according to the main thematic/cross-disciplinary approaches adopted by the programme. A set of potential strategic planning visions, actions, and policies from a WEF nexus perspective is presented, exploiting disciplinary specificities and expertise within a common operative and methodological framework. Territorial readings, guidelines, and scenarios complement and integrate projective and policy-oriented models in suggesting new ways to envision food and energy production and water management, stressing the key role of appropriate cartography in dealing with the complexity of the challenges. This section includes essays by Alice Buoli, Alessandro Frigerio, Davide Danilo Chiarelli, and Lorenzo Rinaldi.

Part 3 (Chapters "Trans-scalar and WEF-sensitive Strategic Scenarios for an Integrated Territorial Development. A Proposal for the Maputo-Boane–Namaacha Transect as a Green-Blue Metropolitan Armature" to "Society: Maputo, a Case of Social Non-simultaneity? A City Repertoire of Issues") presents a collection of relevant lemmas helping in defining the key dimensions of the "urban question" in sub-Saharan African cities. Most of the essays present a general and introductory discussion around the suggested term from a disciplinary perspective and then explore the topic with reference to the specific context of Mozambique and/or Maputo. Each contribution is to be seen as a contextualized/context-sensitive "entry" of an open and transdisciplinary lexicon that complements and integrates the discussion on the past, present, and future of the Maputo Province. This section includes essays by Paola Bellaviti, Paolo Beria, Andrea Castelletti and Elena Matta, Valentina Dessì, Laura Montedoro, Agostino Petrillo, Matteo Rocco, Maria Cristina Rulli, Davide Danilo Chiarelli, Nikolas Galli and Camilla Govoni, Maria Chiara Pastore.

The book is also enriched by a presentation by Emanuela Colombo (Politecnico di Milano Rector's Delegate to Cooperation and Development since 2005) and Manuela Nebuloni (Polisocial Award Organising Committee) and a final contribution by Gabriele Pasqui (Full professor and former director of DAStU, Politecnico di Milano, proponent of the Polimi-UEM memorandum of understanding and PIMI). All the contributions in this book can be seen as autonomous essays, yet we suggest the readers to consider them as pieces of a comprehensive narrative made of different interdependent parts, open to a variety of reading levels and paths. In the multiplicity of gazes and internal references lies one of the main qualities and contributions of this book to international scholarly debates on urban sustainable development (in the Global South and beyond), providing an example of productive interdisciplinary exchange and knowledge co-production across disciplinary fields and research cultures.

References

Behrens R, Watson V (1996) Making urban places. Principles and guidelines for layout planning. UCT Press, Cape Town

Blaustein SM, Goitia C, Mehta G, Plunz R, Pointl, J (2012) Re-cultivating the garden city of Kumasi. Urban Design Lab at the Earth Institute, Columbia University, New York

Botai JO, Botai CM, Ncongwane KP, Mpandeli S, Nhamo L, Masinde M, Adeola AM, Mengistu MG, Tazvinga H, Murambadoro MD, Lottering S, Motochi I, Hayombe P, Zwane NN, Wamiti EK, Mabhaudhi T (2021) A review of the water–energy–food nexus research in Africa. Sustainability 13(4):1–26. https://doi.org/10.3390/su13041762

Broto VC, Ensor J, Boyd E, Allen C, Seventine C, Macucule D (2015) Participatory planning for climate compatible development in Maputo. UCL Press, London, Mozambique

Chiodelli F, Mazzolini A (2019) Inverse planning in the cracks of formal land use regulation: the bottom-up regularisation of informal settlements in Maputo, Mozambique. Plan Theory Pract 20(2):165–181

Colombo E, Pastore MC, Sancassani S (2016) Storie di Cooperazione Politecnica 2011–2016/stories of cooperation at Polimi. Polisocial—Politecnico di Milano, Milan

Colucci A, Magoni M, Menoni S (eds) (2017) Peri-urban areas and food-energy-water nexus. Sustainability and resilience strategies in the age of climate change. Springer, Cham

FAO (2014) The water-energy-food nexus. A new approach in support of food security and sustainable agriculture. FAO, Rome

Gouverneur D (2014) Planning and design for future informal settlements: shaping the self-constructed city. Routledge, New York

Harrison P (2006) On the edge of reason: planning and urban futures in Africa. Urban Stud 43(2):319–335

Hassan Tolba A, Khalifa M, McNamara I, Ribbe L, Sycz J (2018) Water energy food security nexus. A review of the nexus literature and ongoing nexus initiatives for policymakers. Nexus regional dialogue programme (NRD), Bonn

Kurtak K, Daher R (2011) Urban development in Accra, Ghana: an implementation toolkit. Columbia University, New York

Jenkins P (2013) Urbanization, urbanism, and urbanity in an African city: home spaces and house cultures. Springer, Cham

Lawrence R, Despres C (2004) Futures of transdisciplinarity. Futures 36:397–405

Ramadier T (2004) Transdisciplinarity and its challenges: the case of urban studies. Transdisciplinarity 36(4):423–439

Macucule D (2010). Metropolização e Restruturação Urbana: O território da Grande Maputo. Dissertation, New University of Lisbon, Department of Geography and Regional Planning

Macucule D (2016) Processo-forma urbana: Restruturação urbana e governança no Grande Maputo. Dissertation, New University of Lisbon, Department of Geography and Regional Planning

Martin-Nagle R, Howard E, Wiltse A, Duncan D (2011) Conference synopsis. In: Conference Bonn2011 the water, energy and food security nexus solutions for the green economy. German Federal Government, Bonn, 16–18 Nov 2011

Mbembe A, Nuttall S (2008) Introduction: Afropolis. In: Nuttall S, Mbembe A, Appadurai A, Breckenridge C (eds) Johannesburg: the elusive metropolis. Duke University Press, Durham, NC, pp 1–33

Melo V, Jenkins P (2019) Between normative product-oriented and alternative process-oriented urban planning praxis: how can these jointly impact on the rapid development of metropolitan Maputo, Mozambique? Int Plan Stud 26(1):81–99. https://doi.org/10.1080/13563475.2019.170 3654

Montedoro L, Buoli A, Frigerio A (2020) Towards a metropolitan vision for the Maputo province. Maggioli Editore, Santarcangelo di Romagna

Myers G (2011) African cities. Alternative visions of urban theory and practice. Zed Books, London

Pieterse E, Simone, A (2013) Rogue urbanism. Jacana media in association with African centre for cities. University of Cape Town, Auckland Park, South Africa

Petrillo A, Bellaviti P (2018) Sharing knowledge for change. Universities and new cultures of cooperation: transnational research and higher education for sustainable global urban development. In: Petrillo A, Bellaviti P (eds) Sustainable urban development and globalization. New strategies for new challenges—with a focus on the Global South. Springer, Cham, p xiii-xxx

Purwanto A, Sušnik J, Suryadi FX, de Fraiture C (2021) Water-energy-food nexus: critical review, practical applications, and prospects for future research. Sustainability 13(4):1–17. https://doi.org/10.3390/su13041919

Roy A (2009) The 21st-century metropolis: new geographies of theory. Reg Stud 43(6):819–830

Todeschini F (2014) Some reflections on the 'nature of an appropriate and resilient spatial plan' for South African Cities. Paper presented at the UIA international congress, Durban, 2–7 August 2014

UN-United Nations (2015) Transforming our world: the 2030 agenda for sustainable development. United Nations, New York

UN-United Nations (2016) New urban agenda. United Nations, New York

UN-Habitat (2015) Towards an African urban agenda. UN-Habitat, Nairobi

UN-Habitat (2016) International design collaboration for Kenya, competition guidelines. UN-Habitat, Nairobi

UN-Habitat (2017) Implementing the new urban agenda by strengthening urban-rural linkages. UN-Habitat, Nairobi

Wahl D, Ness B, Wamsler C (2021) Implementing the urban food–water–energy nexus through urban laboratories: a systematic literature review. Sustain Sci 16:1–14. https://doi.org/10.1007/s11625-020-00893-9

Watson V, Agbola B (2013) Who will plan Africa's cities? Africa Research Institute (ARI), London

Watson V (2016) Shifting approaches to planning theory: global north and south. Urban Plan 1(4):32–34

Part I
Sustainable Territorial Development of the Maputo Province. Debates, Research, and Innovative Perspectives

A Capital in History: Widening the Temporal and Physical Context of Maputo

Paul Jenkins

Abstract This chapter presents an overview of Maputo in relation to its surrounding region, with an emphasis on spatial and temporal change. It supports an endogenous reading of these scales for what is now one of the top twenty urban conurbations in Sub-Saharan Africa. In doing so it aims to challenge dominant negative discourses of such rapid urbanisation—prevalent in both Global North and South—and stresses the essential dynamism of urban society and culture, albeit within constrained political and economic limits. The chapter is essentially an opinion piece and draws on more than forty years of the author's personal experience, professional work and research in Maputo, and is accompanied by a wide range of the author's prior publications for further reading and reference. Rapid urbanisation in Sub-Sahara Africa, with relatively weak states and still quite limited private sectors, will bring enormous changes in the next few decades to land and environments. The extent of this urbanisation process also means that the alignment of environmental resource management of the "rural" adjacent to these "urban" futures is crucial and Maputo and its surrounding region can be a case study from which much could be learnt for wider trends in Sub-Sahara Africa.

---

[1] As an opinion piece the chapter diverts from conventional referencing format and includes a series of references of general relevance to the material presented, through relevant publications by the author—but does not detail these through the text. Those interested in more detail of other sources can consult the referenced publications, as to list all relevant sources would have extended the reference list excessively.

P. Jenkins (✉)
University of Edinburgh, Edinburgh, UK
e-mail: paul.jenkins@wits.ac.za; p.jenkins@ed.ac.uk

University of the Witwatersrand, Johannesburg, South Africa

© The Author(s), under exclusive license to Springer Nature Switzerland AG 2022   23
L. Montedoro et al. (eds.), *Territorial Development and Water-Energy-Food Nexus in the Global South*, Research for Development,
https://doi.org/10.1007/978-3-030-96538-9_3

1 Introduction

This chapter is primarily a personal opinion piece supporting the intention of this book in reviewing both physical and temporal scales for the city.[1] It emphasises a wide spatial view of the city and its region and also a long historical dimension of urban development. As argued initially some 30 years ago, it is important to have a *"longue durée"* perspective in African urban analysis, as this can counteract dominant Northern negative discourses.[2]

From the author's experience of five decades of living, working and researching in urban areas (of which more than forty years have engaged with Maputo), he has previously criticised the general emphasis of Northern discourses on "disorder" in African urban space and form. This position is primarily as the nature of the prevailing socio-cultural order in urban development in Africa in the post-colonial period has not been dominated by the state or so-called "formal" economies. In addition, the Northern experience of key periods of rapid urbanisation happened in very different political and economic contexts, and the way pre-existing socio-cultural values were impacted by such contexts led to very different outcomes. That is not to say that rapid urbanisation in Sub-Saharan African cities is all beneficial to the majority of urban citizens—far from it—but the socio-cultural response by the majority has been significantly different, and this is directly related to limited state governance capacity and relatively weak "formal" economies.[3]

In general, that which escapes state identification and regulation is usually seen as illegal or "informal", the latter being a term coined in the 1970s, which has expanded in conceptual significance and has come to be used socially and politically in very negative ways. This chapter will use the term "non-formal" for that which the state cannot identify, define, register and regulate (or tax). In most Sub-Saharan African cities this represents the majority of urban land use, environmental access and housing, as well as much infrastructure and economic activity. In this view, there is no clear cut distinctions between so-called "formal" and "informal", as the "non-formal" in fact permeates political, economic, social and cultural life. Maputo is a prime example of this, as this book relates.

This introduction also needs to make reference to limits to the understanding of the *"longue durée"* in African urban analysis. Historiography generally favours the written word over other forms of remembering and as such often displays negative bias in relation to what is broadly termed "pre-history". In the case of African urban areas this means discounting most activity prior to European mercantilism—and even more so in subsequent colonial histories. It is more generally recognised now that urban areas have existed in Sub-Saharan Africa for millennia, but there has as yet only been limited in-depth investigation of the dynamics of these prior to European contact. This has been partly as access to source material is often limited,

---

[2] This "longue durée" approach is discussed in more detail in the author's initial research paper on the city dated 1999 (see above).

[3] The "formal" economy refers to that which is recognised, registered and regulated (to some extent) by the state, otherwise known as the "public" and "private" sectors.

either through records within African urban societies and cultures, or due to lack of interest/funding for deeper investigation (e.g. archaeological). That leaves considerable space for speculation on the African urban past which generally has been negative. However, some trends in pre-history (or at least subaltern trends identified in past written urban histories) are still in fact what positively underpin the dynamics of African cities, which is a key theme of this chapter.

2   Maputo and Its Region in Pre-History

What is now the city of Maputo is in fact part of a wider conurbation which includes the adjoining city of Matola and increasingly the surrounding previously partly rural districts and small towns of Boane (southwest of Matola) and Marracuene (north of Maputo), creating a wide urbanised region which is a de facto metropolitan area, although this has no legal/administrative structure. Maputo was called Lourenco Marques by the Portuguese until Independence in 1975[4] and Matola only developed from the mid 1960s when Portugal began to relinquish some political economic control over foreign investment. Lourenco Marques in its turn was known from (at least) the early colonial period as Xilunguine—or the "place of foreigners" in the predominant local language Xi-Ronga.[5] The Va-Ronga people had occupied the area north and south of Maputo Bay from probably the 9th or 10th century of the Common Era (CE), at which time it appears they migrated from the higher lands to the west—possibly due to over-population. Certainly there is evidence, as yet not fully investigated, of intensive agrarian production and related settlements on the eastern slopes of the higher plateau in what is now Mpumalanga Province of South Africa, with established trade routes including to the "highveld" hinterland as well the Indian Ocean ports further north. These Va-Ronga people, whose descendants, are still widespread in Maputo Province, lived by a relatively settled form of agriculture and environmental management and displaced earlier farmer/hunter/gatherers, as evidenced by early temporary settlements in archaeological excavation in the region dating from the 1st century CE. Those early farmers will have been of so-called Bantu stock, whose migration across to Eastern Africa and then southwards is demonstrated through both linguistic and archaeological evidence.[6]

The relevance here of this quick glimpse of pre-history for the region of Maputo is that the Va-Ronga lived in fairly stable rural settlements of 500 to 2000 people, which had a very decentralised but socio-culturally interlinked polities, and quite advanced economies based on agriculture, mining, manufacture and trade links. These were the peoples described in the first European historical accounts which date from the early

---

[4] The British called the urban area Delagoa Bay.

[5] The chapter uses Maputo for the city whether before or after Independence for ease of reading.

[6] In turn these peoples will have displaced or co-habited with earlier people of Khoisan origins who were the original human inhabitants of Southern Africa (and who had evolved from early hominid roots over millennia).

16th century, the earliest being from shipwrecked mariners. Portuguese mariners were the first to round the Cape and arrived in Maputo Bay by the middle of the 16th century, where they started a rather desultory engagement with the established Va-Ronga—through temporary trading camps—as their primary interest was in finding a route to India as well as the fabled gold wealth of the central African plateau. The Portuguese coming from the South were however not the earliest external contact for the Bantu-speaking farmers, who had long trade links to the coast further north, which in turn had seafaring links into the Indian Ocean Swahili trading culture with Arabia and South Asia, at least from the 9th century CE. The Va-Ronga people were thus well-established in their region which included all of what is now Southern Mozambique, but with wider regional and international linkages.

3   Maputo and Its Region in the Period of European Mercantilism

The Portuguese only established more permanent occupation of Maputo Bay when their trade was contested during the 18th century CE, through increasing competition between European nations, and the Portuguese finally created the first permanent settlement in the 1780s. This initially was partly for fresh supplies for the ships heading to and from the main areas of economic and political interest of India and the Swahili coastal city-states—but also included trade in a range of farmed, gathered, and hunted products for export, the latter especially ivory—as also happened to the south in ports dominated by the English trading with the Nguni hinterland.

In time this wider regional trade in ivory came to dominate other trading interests and Maputo became a key base for elephant hunters venturing far into the hinterland. Competition for this highly profitable resource however led to dwindling animal stocks and increasing contest for hunting control, which in turn has been seen as the factor stimulating friction between and within regional ethnic groups. Historians argue that this then led to the emergence of more centralised forms of indigenous African social and political organisation, especially amongst the Nguni people of eastern Natal, and that this culminated in the Mfecane, or "Zulu Uprising", in the early part of the 19th century CE. However, Nguni states were not the only ones in formation in the wider region in this period, as the Dutch farmers ("Boers") trekked out of their original colonised lands in the Cape north across the Vaal River and established a number of small agriculturally based states in the high plateau in the hinterland from Maputo Bay. The Boers also used Maputo Bay for supplies, leading to more trade going through the small port, at a time of dwindling elephant populations.

The Mfecane directly affected the Maputo settlement and adjoining indigenous peoples several times through invasion—mostly however the Nguni groups passed through on the way north in search of new territory—and the Va-Ronga were to some extent protected by their links with the fragile Portuguese mercantile presence (probably as the port was known as a major trading point for ivory). This socio-political disruption in fact aided establishment of a wider area of Portuguese control (*circa* 10 km radius) around the original small settlement on a sand-bank surrounded by swamps. In general, the early mercantile period of European engagement with the Maputo region thus accentuated and re-oriented previous local and regional trade—especially with the exterior of Africa—and created more instability, one of the results being the large Nguni-based state in Gaza north of Maputo, which dominated many less centralised indigenous groups for a long period.

By the end of the 19th century competition between European powers for natural resources and markets in the rest of the world led to colonisation in many parts of the globe and soon after world warfare between the dominant nations. This process was first manifested in Sub-Saharan Africa by the Berlin Conference of 1884, which divided up the continent into colonial territories, requiring the previous mercantile/protectorate powers to show that they ruled politically and militarily, as the basis for colonisation and hence more systematic and exclusive economic exploitation. That in turn led to colonial investment in railways to aid other forms of natural resources extraction, especially mining and plantation agriculture. In the inner plateau in what is now South Africa the Boer Republics remained essentially agricultural, and together with the Portuguese they planned a railway for export of their produce through what is now Maputo to avoid the British ports. This plan was overtaken however by discovery of diamonds and, soon after, gold in the high plateau Transvaal region, which led to British colonial take-over of the small Boer republics, and the consolidation of what is now South Africa. That in turn spurred on the railway construction from the main mining areas on the South African Rand, and associated construction of the port at Maputo harbour and led to the town being awarded city status and soon after being named the capital of the "Overseas Territory", as Portuguese colonies were called.

In the long mercantile period of slow growth of the small trading town, no "formal" planning was attempted, only military fortification against attack (which proved ineffective). Differently from Spanish rule, Portuguese rule was often rather indirect, and the built environment followed principles of spatial development but no strict regulation or orthogonal planning.[7] This was true for the many Portuguese mercantile settlements worldwide—the majority being ports—and was a flexible approach to quite different physical, economic and social contexts. In Maputo, vestiges of this almost medieval street network (Fig. 1) still remain in the city centre downtown "Baixa" area where the urban area began.

4 Maputo and Its Region in the Earlier Colonial Period

The small "organic" urban fabric of early Maputo changed markedly with colonial possession and associated parastatal investment in transport infrastructure. The

---

[7] As was the case of the Laws of the Indies in settlements established in the New World by Spain.

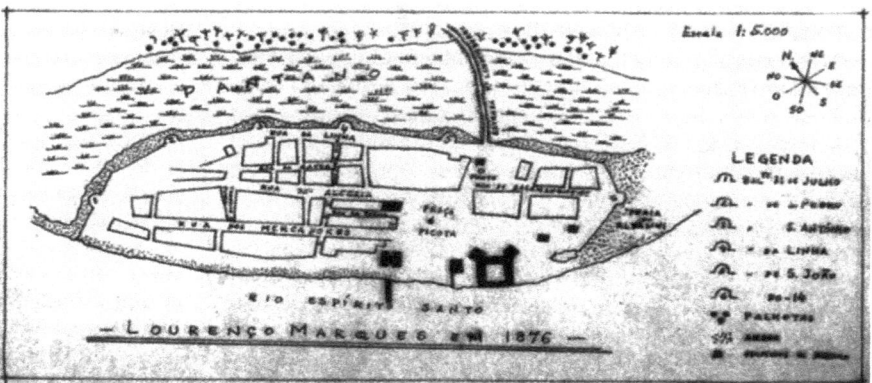

Fig. 1 Plan of Maputo in 1876 showing ramparts, fort, customs house, open plaza and governor's house—the key elements of Portuguese mercantile urban planning. *Source* author's personal archive

extensive new port area, with its highly decorative main railway station,[8] as well as a number of neo-classically designed public buildings, sprang up rapidly in the centre, and a new orthogonal urban plan was laid out crossing (and draining) the surrounding swamps and moving up the escarpment to higher ground. These rather primitive initial urban plans were as much about engineering as spatial planning and permitted basic urban infrastructure to be implemented (albeit usually initially with foreign capital). However, the move up from the low lying "Baixa" centre meant encroachment on indigenous land, which was already interspersed with foreign land-holdings, some of the occupants of which had an eye to speculative gain seeing the town expand.

The Portuguese state began consolidating its control by expropriation of indige-nous land and buying back other land from foreign landholders, to be able to insti-tute land leasing arrangements by the Crown (a slow process resolving existing land ownership disputes, which started from 1858). The first formal state land register was created only in 1886 in fact just before the town was raised to city status, with the first full urban plan soon after in 1896. In addition, the new urban plan had to engage with incorporating the separate "English township" already established on the Polana headland, created by the strong British interests in *inter alia* urban infras-tructure investment (e.g., telegraph, gas and water supply). In this process of state consolidation and centralised planning of land, indigenous land rights were removed in 1890, except in urban 'reserves' (which were only created from 1918). These reserves were located outside the new city land registry area—and predominated to the northwest as much land to the northeast was still in private landholding(s) where a few foreign speculators resisted state acquisition—these legal disputes lasting until mid 20th century.

---

[8] The design often mistakenly attributed to Gustavo Eiffel.

The urban economy radically changed with the disappearance of the individually organised hinterland hunting and ox-cart based trade, largely due to the development of new state-dominated railway/port traffic (and associated administrative structures), as the basis for new forms of national and macro-regional trade. However, there was very slow physical development of the planned settlement until well after trams were installed in the 1920s. Urban settlement also expanded non-formally in a significant way to the northwest outside the planned and registered urban area from at least the 1930s, with multiple small colonial land concessions, mostly used for indigenous African housing through land and/or shack rentals. These were legal land holdings and in fact the building regulations ended at the edge of the so-called "cement city", and they provided shelter of an extremely basic nature, much of which was actually self-built by the occupants. Outside of the core "cement city" the state in fact had limited interest to control any development and used the indirect form of colonial regulation through "native chiefs" called *regulos* in the "native reserves" and the new peri-urban areas to the northwest. Although these *regulos* had considerable local power it was always circumscribed and dominated by the colonial state, which also kept municipal government weak to avoid settler interests conflicting with national interests (Fig. 2).

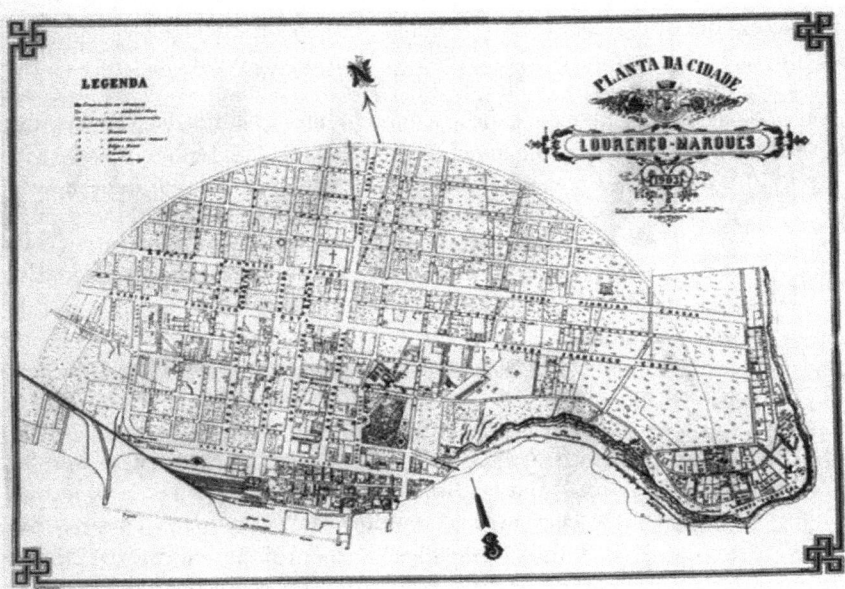

Fig. 2 Plan of Maputo in 1903 showing the land registry area, now also incorporating the pre-existing "English township" on Polana headland to the southeast. Note also the port and railway development to the southwest. *Source* author's personal archive

5   Maputo and Its Region in the Middle Colonial Period

The early and middle colonial periods were in fact relatively short compared to the much longer mercantile period, and it was several decades into the 20th century CE before Portuguese rule over the whole national territory was established. In the interim large parts of the north and centre of the country were leased to foreign companies with extensive governing powers. In Maputo the city only began to infill the initially orthogonally planned city council area by the 1940s, by which time the non-formal areas to the northwest were already well established. Quite unlike the trend in surrounding British colonies—and despite the existence of racially-based legal/administrative separation of opportunity—there was no formal racial separation in urban space—albeit there was definite class separation. In the non-formal areas, the form of socio-cultural mixing that had characterised the earlier mercantile period continued to a significant extent but was gradually ground down with growing colonial settlement of poor peasants from the metropole (as a way to allow land consolidation and nascent industrial development in Portugal). The state in Mozambique continued relatively weak—central government as much as local government—and the economy was partly predatory on the indigenous population and natural resources, and partly subordinate to the growing mining industrial complex of its powerful neighbour South Africa.

By the middle of the 20th century colonial settlement schemes and indigenous labour migration had underpinned rapid urban expansion in Maputo: the former trickling down to the 'cement city' with growing class and racial division, and the latter evident in non-formal settlements, mostly through rented land and/or housing from settler landowners. The city economy continued to be trade-based, mostly focussed on export of un-processed natural products from obligatory peasant farming in the north, colonial plantations in the centre, and other colonial farms in the southern region of the country (e.g., sugar, cotton, cashew nuts etc.)—but with growing importance of the administrative and financial sectors. During the Second World War Portugal remained neutral and as a result this stimulated incipient industrialisation of some local produce due to global shortages—Maputo developing a small secondary industry sector to the west of the "Baixa", and around the large non-formal areas of Xipaminine/Chamanculo to the northwest. Industry remained limited, but the indigenous population in the capital grew significantly, always outnumbering the increase in colonial settlers.

Local government investment was primarily for the "cement city", which slowly developed with southern European styles of urbanism (e.g. tree-lined avenues) and architecture (e.g. Art Deco villas). A separate housing scheme was built for 'assimilados' in the 1940s—small in scope and relatively expensive—but it did nothing to redress increasing urban spatial imbalances. Some new bourgeois and petit bourgeois satellite settlements were also developed in this period (e.g. in Machava—where another industrial area was started). In this context an urban master plan for Maputo was prepared by the Portuguese colonial planning office (published 1952), however this avoided most of the northwest non-formal settlements and focussed on expropriation from the foreign land speculators to the northeast. The urban boundary and the land registry area in Maputo were extended significantly to the north in 1965 by when 88% of the 770 Ha of administrative urban area was in private freehold/leasehold, 85% of which being held by 11 landowners—clearly demonstrating the local concentration of capitalist wealth created by the colonial fascist regime—soon to change.

In this middle colonial period, the city began to expand into its immediate region and beyond its new municipal boundaries, both formally with new planned satellite areas, and non-formally in "native reserves". The region around the growing urban area also changed economically with more intensive agriculture in sponsored irrigation areas in Maputo Province, partly driven by growing urban demand (and also the disruption of the IInd World War) and often as state planned development projects. But Portugal was also changing—the long-term dictatorship went through a period of challenge and revision, and this led to a change of emphasis from closed national economic development, closely associated with the fascist regime, to a wider opening to global capitalism, albeit very defensive in aspiring to retain colonial "possessions". This was very different from other European colonial nations such as France and Britain, which were only too happy to relinquish territories to post-colonial elites due to post-war economic exhaustion and also US pressure to permit more open global economies (Figs. 3 and 4).

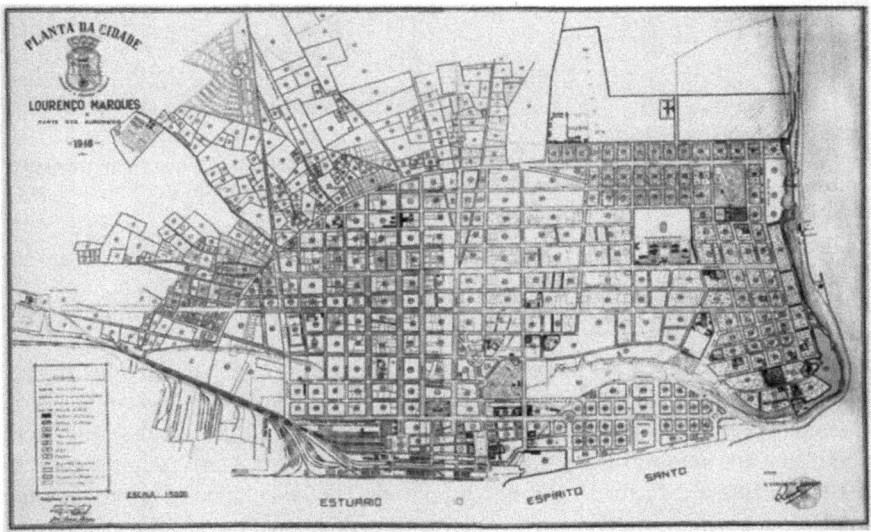

Fig. 3 Land cadastre plan of the city in 1940 showing the unplanned expansion to the northwest and also the planned area for 'assimilados' north of this. *Source* author's personal archive

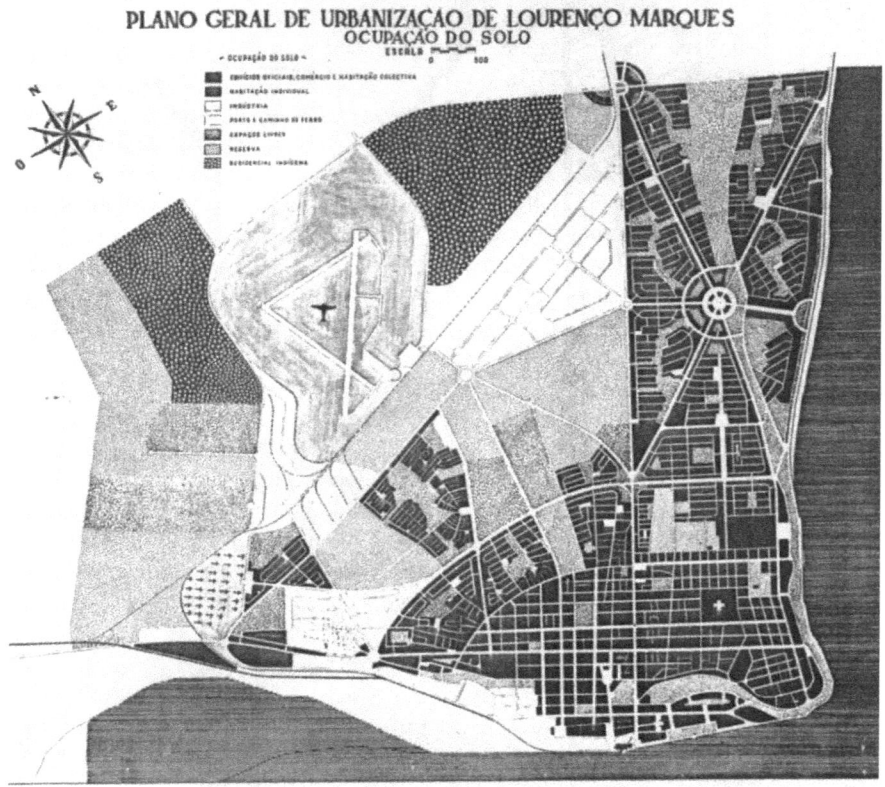

Fig. 4 Maputo general urban expansion 'Aguiar Plan' of 1952, showing the primary planned expansion to the northeast, avoiding most existing unplanned 'reserves' to the northwest. *Source* author's personal archive

6   Maputo and Its Region in the Later Colonial Period

The later dictatorial regime in Portugal was influenced by global capitalist trends, which led to some changing city-region relations in Maputo, with less emphasis on dominant Portuguese investment (often in rural and agricultural sectors) and more on industry and mining, as well as major national infrastructure projects such as the Cabora Bassa dam for hydro-electric power. This also came at a time of (and stimulated even more) rapid urbanisation, with wide impact on natural resources in the city region, as well as the previous indigenous urban land access patterns, which had been largely non-monetary (except for "gifts" to *regulos*). The new colonial administration published a number of national development plans, and for Maputo it also started a more comprehensive physical planning process in 1969, not only based on future designed physical aspirations (such as the plan from the 1950s), but with a more empirical base in social and economic trend analysis. However—although the resulting urban "Plano Director" (prepared by Portuguese consultancy) did incorporate a wider urban regional vision (which a local planning office had initially drafted a short time previously) —it still focussed primarily on the city municipal area and was still a master plan. As such, it only projected a desirable physical future and did not engage adequately with the political economy of urban growth and socio-cultural trends. Nevertheless, it was much more "modern" in intent, paralleling the modernist trend in "Tropical Modernist" architecture of the 1960s, which began to challenge the proto-fascist "Soft Portuguese" architecture style that had been state-sponsored from the mid 1940s (Fig. 5).

The modernist vision that this urban plan proposed (similar to the superficially utopian modernist output of contemporary architects such as Pancho Guedes) never engaged with the long-term *realpolitik* of a weak governance regime—which

Fig. 5 Urban "Plano Director" of 1972, detailing planned land uses for the city-region. Note the proposed bridge across the estuary (to the west, from Matola). *Source* author's personal archive

included very limited land and environmental control—while implicitly (and some-times explicitly) criticising the state for permitting such development. The new urban masterplan for instance largely ignored the complexity of the long-existing non-formal areas to the northwest and concentrated on the development of newly state-expropriated land to the northeast, destined for new middle-class housing as well as new industrial areas (e.g. what is now Avenida das Forças Populares de Liber-tação de Moçambique). The opening of the economy to global investment did not however benefit Maputo, as much of this went to the new municipal area to the west in Matola, where a large scale new industrial port and associated factories sprang up quickly in the 1960s, accompanied by new housing. Here the municipality was more enlightened in developing low-cost land and housing support for the emerging industrial workforce—unlike in Maputo where such assistance was extremely limited (e.g. a small low-cost modernist extension of minimalist apartments in Malhangalene neighbourhood). The industrial area created along the railway line to South Africa in Machava also developed rapidly in this period and the local authority there, as in Matola, developed some low-cost sites and services housing schemes also for indigenous Mozambicans.

It was only with the growing impact of the Independence war in the late 1960s and early 1970s that the central state changed its tactic in relation to the wider indigenous population in Maputo, with a few sites and services schemes planned (albeit only implemented after Independence). The growing impact of the war for liberation in the colonies had a significant impact on Portugal, as did the emerging possibilities of Portugal seeking a more beneficial future role in Europe as opposed to relying on its colonies. This—together with international pressure against obvious colonial discriminatory practices and widespread resistance by Portuguese to conscription for colonial wars—led to revolution in Portugal in 1974 and the rapid handover the following year of Mozambique to the only viable independence movement, a front of various previous organisations called FRELIMO. Unlike in the managed decoloni-sation of other African territories, this hand over was abrupt and more conflictive, partly of course due to the decade-long military struggle for Independence, which in turn politically radicalised the Independence movement. That radicalisation however was also due to the attitude of the major global powers in the ongoing Cold War—with Mozambique being caught as a proxy between East and West—something that dogged independent development for years to come.

7   Maputo and Its Region in the Immediate Post-Independence Period

In many ways Independence (including its immediate antecedents and consequences) brought significant changes to Mozambique and also to its capital city Maputo, which became governed by a central government appointed Executive Council and incorporated the previously separate Machava and Matola urban areas. All land in the country was nationalised, as was rented and abandoned housing—which was primarily in the capital's conurbation. Many of the Portuguese colonial population and other sympathisers abandoned the country, and this had a marked effect on social and economic life in urban areas—too many to detail here. One unforeseen impact however was the many residential mortgages from the 1960s building boom that were also abandoned, and thus represented significant losses for banks which were also intervened by the state. While the impact of Independence was probably more direct in the central "cement city" of Maputo it also impacted on the "suburbios"— the long-existing non-formal areas surrounding the cement city—where much land and housing had been rented from settlers, this also being nationalised. Another important socio-economic impact on the conurbation was a rapid increase in urban influx as families joined existing (often male) members who already worked in the city, which led to widespread pressure on land in peri-urban areas surrounding the existing formal and non-formal occupied urban areas.

FRELIMO was a relatively weak political force—at least initially—although it transformed itself rapidly from a liberation guerrilla army into a government administration. One way the single party and government managed urban life was through creating neighbourhood "dynamising groups" (Grupos Dinamizadores, GDs) which handled local-level administration and had clear political functions passed down from central government.[9] Given the FRELIMO adversity to the previous "traditional" secondary governance system of colonially appointed *regulos*, the GDs played important roles in urban land access, although this was not officially their responsibility. So, when the City Executive Council developed its management structure in the early 1980s, creating a department to deal with urban planning and construction (Directorate for Construction and Urbanisation—DCU), it was in conjunction with these GDs that the DCU was able to survey, plan, allocate and guide land use and basic house construction in a new strategy which focussed on sites with very basic services with support to self-managed house construction, overtaking the limited pre-Independence initiatives in this field. The DCU in fact had been expected by the new National Housing Directorate to take over a pilot neighbourhood upgrading project in the city (Maxaquene/Polana Canico), which had been funded by the United Nations—but it opted for a wider urban strategy of creating a belt of basically planned land around the (then) perimeter of urban occupation in Maputo, Machava and Matola to guide on-going expansion trends—with considerable success despite the limitation in funds and capacity. In this, it built on the pre-existing indigenous traditions of land access in the peri-urban area.

In fact, although all land and some housing had been nationalised, the FRELIMO government had no urban policy, focussing instead on large-scale agricultural and infrastructure investment, stimulating basic industry, and attempting a proto-socialist form of centralised development—and in this it was supported by the Eastern Bloc.

---

[9] These were arguably stimulated by the experience of grassroots community organisations which arose in Portugal in the revolution (with similar names), but in the Mozambican case, were state-instigated and dominated. However, they represented a very important interface between the state and society.

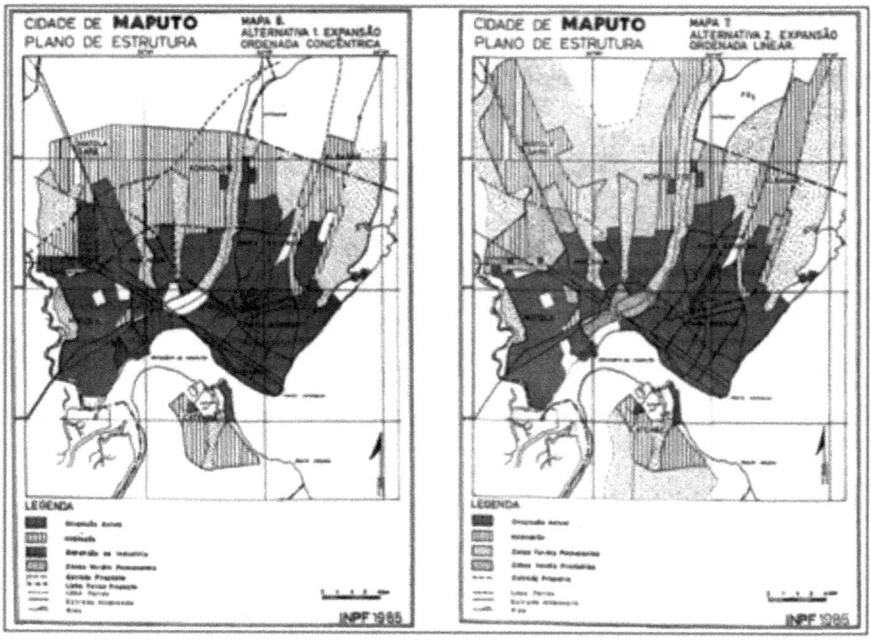

Fig. 6 City Structure Plan 1985 showing two possible planning alternatives for the wider city area: one a high investment concentric option (left); and the other a lower-cost investment linear option (right). Neither was effectively implemented. *Source* author's personal archive

In general—perhaps because of the rural-based experience in the guerrilla war—FRELIMO tended to see urban areas as parasitic, which eventually led to attempts to forcibly remove unemployed and "marginal" people in the "Operation Production" of 1984—somewhat similar to forced villagization in (for example) Tanzania. This failed, and in the process it nearly bankrupted the national airline and turned many against FRELIMO—and badly undermined the legitimacy of the GDs which had the final decision on who was removed. This implicit anti-urban bias of the central government was reflected also in investment and administrative capacity-building, with the small National Housing Directorate (DNH) mostly taken up with national level projects such as new industrial and linked housing locations as well as communal village creation. However the DNH did undertake urban and regional planning and crucially reviewed previous colonial development plans, for example in Maputo Province. In time (mid 1980s) it also supported local Executive Councils in urban planning and produced a structure plan for the conurbation of Greater Maputo in conjunction with the city DCU. Importantly the DNH trained many middle level planning technicians, who played key roles as intermediaries in a wide range of development projects with their basic physical planning knowledge[10] (Fig. 6).

---

[10] The DNH was transformed into the National Physical Planning Institute (INPF) in the mid 1980s.

8 Maputo and Its Region in the Transitional Post-Independence Period

The proxy Cold War took a terrible toll on Mozambique in general, with firstly Zimbabwe and then South Africa being supported by the Western Bloc to wear down perceived communist influence in the country.[11] This manifested itself in blockades on the economy and essential supplies, direct military aggression, and (most dangerously) indirect guerrilla insurgency against the new government. Despite a formal Non-Aggression Pact with South Africa in 1984, the insurgency peaked in the late 1980s and early 1990s, and in this period the Maputo conurbation was the destination for large numbers of people from its immediate surrounding region (Maputo and Gaza provinces especially), as well as displaced people nationwide, and the city periphery was attacked almost every night, causing a major population contraction, and densification of existing peri-urban areas. However, after peace was negotiated in 1992 the opposite occurred, urban expansion recommenced with renewed pace at a time when local government was in disarray. That disarray was partly due to the reduced availability of government funds for any development projects but also a hollowing out of government capacity when the local currency was multiply devalued overnight. This was the impact of macro-economic restructuring demanded by international agencies, as civil servants sought alternative employment/incomes—those who could, did so in the fast growing international aid sector. In parallel urban land rose in real value despite its officially free cost, a process initially instigated by the GDs, as they became the main way to access land non-formally due to government contraction in official urban planning capacity, but soon also illegally through the few formally planned and allocated land projects, as other civil servants resorted to corrupt practices.

In summary, concerning the physical context for Greater Maputo, this period of abrupt change saw rapid city growth vis-à-vis the surrounding regional population as previous economic activities in the province (and wider rural areas nationwide) contracted. This was accentuated by the approaching civil war, which led to densification and a temporary reduction in urban expansion, but the emergence of (initially hidden) commercialisation of urban land. Social and economic activity in the Maputo city and its region was significantly disrupted by the war, while more and more urban residents entered non-formal commerce as the only economic alternative. The advent of peace, soon after the severe macro-economic devaluation, led to even more open commercialisation of urban land, with the city authority and the theoretically subordinate local neighbourhood administrations (now changed in name from GDs) playing key roles in this de facto decentralised land access process. Meanwhile, central government revised its mainstream development focus away from the proto-socialist ideals of (state-led) agriculture and industrialisation to revert to the late

---

[11] For these neighbouring countries it was part of an attempt to withstand decolonisation (Zimbabwe) and democratisation (South Africa).

colonial strategy which had focussed on natural resource extraction through large-scale foreign direct investment, as indeed many other countries did in the post-Cold War globalisation of neo-liberalism (Fig. 7).

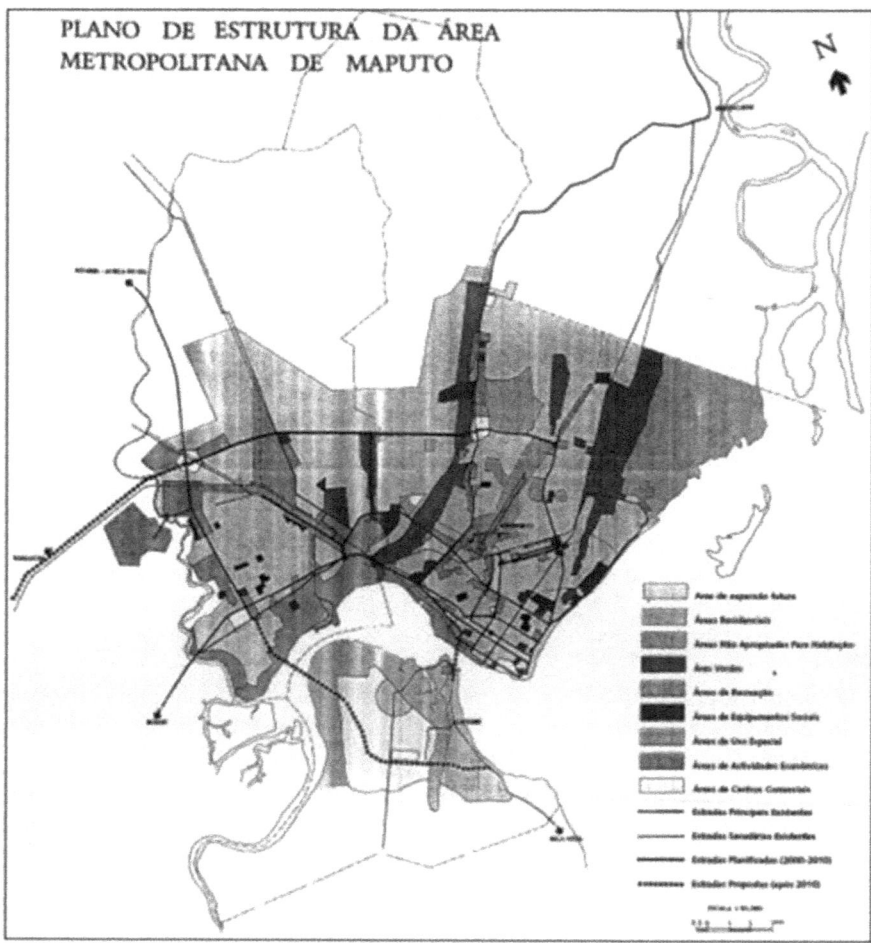

Fig. 7 Greater Maputo Structure Plan 2000: main land use and infrastructure proposals (financed by the World Bank: not approved). *Source* author's personal archive

9 Maputo and Its Region in the More Recent Post-Independence Period

The new change (arguably reversion) of *realpolitik,* manifested from the mid 1990s, saw continued acceleration of urbanisation nationwide, including most obviously in the largest conurbation based on Maputo—which with new local government legislation had been divided into two municipalities—initially Maputo and Matola cities (the latter subsuming Machava) and soon Boane (Marracuene soon due to also be an independent municipality). The urban expansion of the 1990s (predominantly non-formal) continued with some new attempts at formal residential planning in the northern parts of Maputo and Matola. However, urban expansion soon passed existing city limits—especially to the north towards Marracuene and to the southwest towards Boane along existing national roads—and then also to the northwest of Matola, with new state investment in the Mozal aluminium factory and the new EN4 motorway to South Africa. While some people reverted to living and working in the Maputo Province, many more became urban commuters with longer range daily bus/mini-bus and train traffic and the wider Maputo province became even more dominated by the conurbation. The World Bank funded a metropolitan structure plan in the late 1990s, which attempted some integrated planning of the new urban expansion, but this was never approved. Danish aid also attempted a metropolitan environmental management plan early in the new millennium, but this was never finalised. By the end of the first decade on the millennium two new structure plans were eventually approved for the municipalities of Maputo and Matola, but with poor linkages between them, and none with the surrounding de facto metropolitan expansion (Fig. 8).

A new trend in the 2000s was the growth in the local middle class (which had always existed but been reduced by settler exodus and had grown slowly under proto-socialism), but this had picked up momentum in the transitional period. The growing evidence of differential socio-economic patterns soon became widespread, no more so than in new housing areas targeted exclusively to the middle class—some gated and others not—as well as socio-economic change within existing residential peri-urban areas (to some extent through gentrification, but also household wealth acquisition). Recent research has shown that this has been a de facto, if not official, local government strategy (and possibly an implicit central government policy). Local municipalities and administrations deliberately planned much wider areas than previously, with middle class occupants explicitly in mind, as this was also a source of municipal income. The locations of such areas preferentially favoured good locations (e.g. along the Maputo coast) and main transport routes (e.g. along the EN4 out of Matola), and the enormous influx of second hand cars imported by the growing middle class led to wider commuting (and significant traffic problems in peak hours due to inadequate road infrastructure).

A government response to this was central state investment in large-scale road infrastructure, with a new Ring Road around the conurbation (and linking roads to the main economic areas), and a new bridge over the bay to the as yet underdeveloped

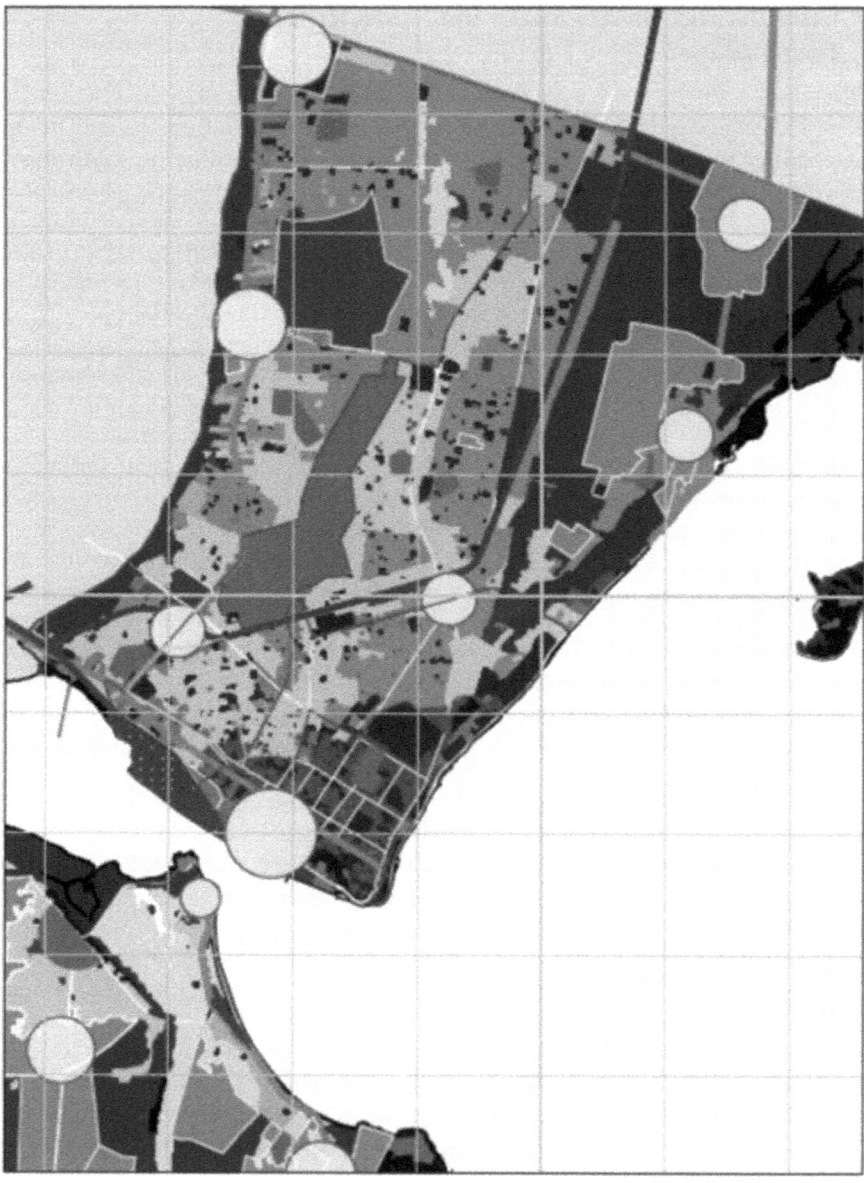

Fig. 8 Maputo Municipal Structure Plan 2010: main land use and infrastructure proposals *Source* author's personal archive

area of Catembe. Catembe had been defined as part of Maputo city from 1980, but the deficient ferry boat service had led to its minimal development despite it being immediately across a narrow part of the bay from the Maputo city centre "Baixa".

The government announced a "new city" to be developed in Catembe—and while this was meant to be for all income groups (as proposed in the 1990 new structure plan as developed by the main state university)—it was clearly predominantly for the middle class.[12] Meanwhile the new Ring Road not only helped the middle class who were hollowing out the cement city as they sought larger residential spaces (and often multiple plots as speculative investment), in which they were quickly followed by commercial and social services—but the road also provided new options for public transport and effectively opened up whole new areas for urban development for the general city population, mainly in the north of Machava. Recent research of middle class development in the city region has shown that this has spread much further than the existing two main municipalities and stretches far into, and even beyond, Marracuene district and Boane municipality.

In all of this, the government still has no explicit urban policy, and there has been an urgent need to review the structure plans for Maputo and Matola for some years (not to mention plan for the wider de facto metropolitan area). While central government is still dominated by FRELIMO, it faces stronger opposition, and some other political parties have gained power in local municipalities nationwide, also seriously challenged the ruling party in elections in the conurbation. This has led to a lag of interest from the central government for any further decentralisation, but also a vacuum in strategic urban direction. Urban expansion remains driven by growing social and economic demand, with limited state control, and in fact the emergent commercialised processes of local access to land have become even more widespread. In addition, since the initiatives by the DNH in the mid 1980s, there have been no real attempts at cohesion in physical and/or environmental regional planning to include municipalities (whose plans stop at their borders), as well as other rural authorities e.g. in the wider city region. What might this lead to is a question the following section discusses (Figs. 9 and 10).

10   Maputo and Its Region in the Future

There is a saying that what we learn from the past is that we never learn from the past…. and that is quite possibly true for this brief overview of urban spatial history. This concluding section presents alternative projections for Maputo for the not very distant future (*circa* 10 *years* perhaps), and as often happens, these are threefold: one which is "ideal"; one which is to be avoided; and one which is pragmatically recommended. However, the challenge of projections is how such analyses can have influence—and experience tells that these seldom have impact directly on policy (as *realpolitik* takes significant time to change), but more likely could have some effect

---

[12] Development of this area is still in process, but much slower than projected.

Fig. 9 Recent research showing new planned areas for the middle class—reaching into the province. *Source* Vanessa Melo & Paul Jenkins 2020

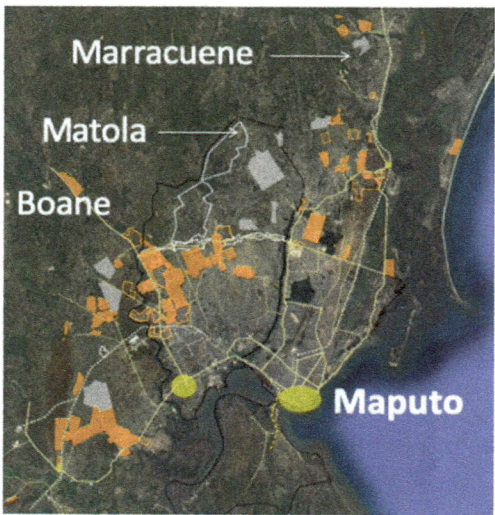

Fig. 10 Maputo in its immediate region—showing urban expansion into the province to north and southwest, also starting to northwest—along main road infrastructure. *Source* Map data ©2021 Google

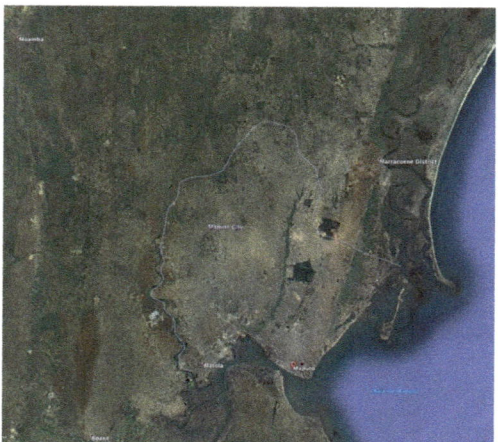

indirectly through education and information provision in its widest (public) sense—at least in the medium term. However, above all, the essential issue is how to link "physical planning" with urban *realpolitik*, including existing forms of governance in situ, and the real "non-formal" urban economy and socio-cultural trends (e.g. in land access and development).

The three alternative scenarios offered are: (1) renewal of attempted top-down land and environmental control, as in current state-based *de jure* approaches; (2) slow and continued slippage to "laissez faire" praxis, as the de facto norm in most recent times; and (3) a shift to enlightened state "guidance" through decentralised planning and public–private partnerships in infrastructure and other urban investment as a better pragmatic option, importantly to include specific state support for private social/non-governmental roles.

The first will fail in the author's opinion, as every attempt at centralised/"top-down" land and environmental control since the end of the mercantile period in Maputo and its region has failed to manage urban growth, whether initial "hard" or later "soft" colonialism and also initial post-Independence proto-socialism, transitioning to the neo-liberalism of recent times. In the brief post-Independence period there was some attempt at aligning local planning with socio-cultural norms and the *realpolitik* of the time, but central government (with its strong control mindset but limited funds and capacity) never adequately supported this. Otherwise the state has always assumed it can and will make the city in its aspirational image—where physical order is aligned conceptually with political administrative order. In reality one form or other of de facto decentralised land and environmental access—albeit without much strategic direction—has prevailed under these various regimes throughout urban history, as this chapter has outlined.

The second alternative—of slipping into a weakly controlled neo-liberal regime (where land rights are quite possibly privatised again in some form) —is the main identifiable current trend, but this is likely to mean less strategic direction, as the formal market and middle class become dominant, and experience elsewhere (e.g. Brazil and Kenya) shows this has long-term negative impacts on socio-economic differentiation and resource management. In this scenario the urban area will continue to expand and regional development would also be predominantly market-driven and more explicitly exploitative (e.g. more obvious contestation for bulk water supply and increased challenges for coastal management to mention a few issues). In this scenario, the elite (who already access most of what they want) and the middle class come to use "rights to the city" to their increasingly exclusive advantage, in conflict with the rights of the majority (which were briefly a dominant strategic objective post-Independence).

The third scenario tries to draw on the historical and physical *longue durée* outlined here, basing itself on the *realpolitik* of the current status (i.e. weak local government, growing middle class and strong private sector interests in public policy formulation etc.). It recognises the potential political gains of developing wider low-cost urban social programmes in land and environmental management—necessarily seen as being decentralised to distinguish effective demand from idealised basic needs. This could be based on an endogenous form of planning—which differs from early indigenous resource management such as in the pre-historical and mercantile periods when the state was weak and recognised this fact. It would however need investment in basic education of citizens, and particularly neighbourhood administration—as well as education and activation of a renewed middle technical level of "grassroots" planners to work directly with communities. It would also benefit from appropriately sensitive participatory budgeting—and the stimulation of a new form of urban actor in the sector: the private social enterprise/local non-governmental organisation (through e.g. tax breaks, facilitated government partnerships in bureaucracy, as well as some financial guarantees). This chapter argues that this form of land and resource management has roots in pre-history and throughout the city's development and has positive potential to be integrated into urban management.

11 Concluding Remarks

In concluding, the central government has a very important role in national urban policy, and local government has such a role in strategic urban direction (e.g. at structure plan level), but urban administrative districts need to deal with land use and environmental planning on a more detailed on-going basis within these policies and strategies (ideally with assistance from non-governmental and/or social agents), and the private sector needs a supporting and investment role at each level (but not a dominant one). Without this shift of direction to a state role of guidance from one of (attempted) control, the capacity to plan and manage limited land and infrastructure for the conurbation, as well as regionally significant resources (environmental as well as economic), will continue to remain side-lined, and gains of Independence for the majority of Maputo's population stand to be lost.

Rapid urbanisation in Sub-Sahara Africa, with relatively weak states and still quite limited formal private sectors, will bring enormous changes in the next few decades to land and environments. The Maputo conurbation will inevitably continue to expand (and densify)[13]—and require more environmental support from its region—and this needs policy, strategy and integrated planning based on local interests, to which this book intends to contribute. The extent of the urbanisation process also means that the alignment of environmental resource management of the "rural" adjacent to these "urban" futures is crucial, and Maputo and its surrounding region can be a case study from which much could be learnt for trends in wider Sub-Sahara Africa.[14]

References

Jenkins P (1999) Maputo city: the historical roots of under-development and the consequences in urban form. Edinburgh College of Art/Heriot-Watt University, School of Planning & Housing, Research Paper 71

Jenkins P (2000a) Urban management, urban poverty and urban governance: planning and land management in Maputo, Mozambique. Environ Urban 12(1):137–152. https://doi.org/10.1177/095624780001200110

Jenkins P (2000b) City profile: Maputo. Cities 17(3):207–218. https://doi.org/10.1016/S0264-2751(00)00002-0

Jenkins P (2001a) Strengthening access to land for housing for the poor in Maputo, Mozambique. Int J Urban Reg Res 25(3):629–648. https://doi.org/10.1111/1468-2427.00333

---

[13] More through natural demographic growth than in-migration in fact.

[14] As noted at the start of the chapter, these publications cite an enormous number of other sources—which it not possible to include here—but which readers can investigate if interested.

Jenkins P (2001b) Emerging land markets for housing in Mozambique: the impact on the poor and alternatives to improve land access and urban development - an action research project in peri-urban Maputo. Edinburgh College of Art/Heriot-Watt University, School of Planning & Housing, Research Paper 75

Jenkins P (2003) In search of the urban–rural frontline in postwar Mozambique and Angola. Environ Urban 15(1):121–134. https://doi.org/10.1177/095624780301500115

Jenkins P (2004) Querying the concepts of formal and informal in land access in developing world—case of Maputo. In: Vaa M, Hansen K (eds) The formal and informal city—what happens at the interface. Nordic Africa Institute, Uppsala

Jenkins P (2006) The image of the city in Mozambique. In: Bryceson D, Potts D (eds) African urban economies: viability, vitality or vitiation of major cities in East and Southern Africa? Palgrave Macmillan, London

Jenkins P (2008) Investigating the concepts of legality and legitimacy in sustainable urban development: a case study of land use planning in Maputo, Mozambique. In: Sassen (ed) UNESCO encyclopaedia of life support systems volume on human settlements and sustainability (EOLSS), vol 4. EOLSS Publishing Ldt, Oxford, pp 163–183

Jenkins P (2009) African cities: competing claims on urban land. In: Nugent P, Locatelli F (eds) African cities: competing claims on urban spaces. Brill, Leida

Jenkins P (2011) Xilunguine, Lourenço Marques, Maputo—structure and agency in urban form: past, present and future. In: Bakker K (ed) Proceedings of the African perspectives 2009. The African Inner City: [Re]sourced. University of Pretoria, Pretoria

Jenkins P (2012) Lusophone Africa: Maputo and Luanda. In: Therborn G, Bekker S (eds) Capital cities in Africa south of the Sahara. Human Science Research Council, South Africa

Jenkins P (2013) Urbanisation, urbanism and urbanity in an African city: home spaces and house cultures. Palgrave Macmillan, USA

Jenkins P (2017a) Working with urban expansion and densification in Sub-Saharan Africa: learning from land access and urban development in Maputo. Trialog 120:28–34

Jenkins P (2017b) Towards a better understanding of rapid urbanization in Sub-Saharan Africa. In: Feijó (ed) Movimentos migratórios e relações rural-urbanas: Estudos de caso em Moçambique, Alcance Editores, Maputo

Jenkins P (2021) Colonial and post-colonial continuities and discontinuities in urban infrastructure in Africa: a case study in Maputo. In: Vaz Milheiro A (ed), Coast to coast—late Portuguese infrastructural development in continental Africa (Researchers' book). AMDJAC, Porto

Jenkins P, Wilkinson P (2002) Assessing the growing impact of the global economy on urban development in South African cities: case studies of Maputo and Cape Town. Cities 19(1):33–47. https://doi.org/10.1016/S0264-2751(01)00044-0

Jenkins P, Smith H (2003) Collaborative knowledge development through action research - a case study of a research approach: appropriate land management mechanisms for peri-urban areas in Mozambique and Angola. Int Dev Plan Rev 26(1):121–134. https://doi.org/10.1177/095624780 301500115

Jenkins P, Anderson JE, Nielsen M (2015) Who plans the African city? A case study of Maputo: Part 1—the structural context. Int Dev Plan Rev 37(3):39–350. https://doi.org/10.3828/idpr.201 5.20

Jenkins P, Mottelson J (2020) Understanding urban density in Africa, where unplanned and non-formal settlement dominates: a case study of peri-urban Maputo, Mozambique. In: Harrison P, Todes A (eds) Density in Africa. Elgar Publishing, Cheltenham

de Pacheco MV, Jenkins P (2019) Between normative product-oriented and alternative process-oriented urban planning praxis: how can these jointly impact on the rapid development of metropolitan Maputo, Mozambique? Int Plan Stud 26(1):81–99. https://doi.org/10.1080/135 63475.2019.1703654

de Pacheco MV, Jenkins P (2021) Peri-urban expansion in the Maputo city region: land access and middle-class advances. J South Afr Stud 47(4):541–565. https://doi.org/10.1080/03057070. 2021.1939499

Nielsen M, Jenkins P (2021) Insurgent aspirations? Weak middle-class utopias in Maputo, Mozambique. Crit Afr Stud 13(2):162–182. https://doi.org/10.1080/21681392.2020.1743190

The Demography of the Maputo Province

Inês Macamo Raimundo

Abstract The Maputo Province is located in southern Mozambique, bordered by the Republic of South Africa and the Kingdom of e-Swatini, a key geographical condition among the reasons for fast demographic growth, side by side with higher job opportunities internally and regionally along with the proximity with Maputo city, the capital of the Republic of Mozambique. At the same time, internal and international migrants either use the Province as the final destination or as transit to access the neighbouring countries. This chapter is drawn on readings and published articles on different aspects of migration and population studies in Mozambique and Maputo, based on the long-term research trajectory of the author. Available data illustrate the demographic dynamics of the Province from the first post-independence census held in Mozambique in the year 1980 up to the most recent of 2017. The trends emerging from them demonstrate that Maputo Province is not far from the so-called "immigration crisis threshold" as the faster growth has pressurized local infrastructures and is a gate for irregular migration.

1 Introduction

Data from the third and fourth censuses of Mozambique realized respectively in 2007 and 2017 demonstrate that Mozambican population increased in all provinces, but unevenly. Interestingly the City of Maputo, the capital of the Republic of Mozambique, registered a decline of its population size at an order of 11.3%, while Maputo province increased by 58.3% (CAP 2020). However, the reasons that explain the decline of the population of Maputo and the increase of the population of the Province of Maputo are still part of speculation or hypothesis that can be related to:

1. The increase of the land market (Kihatos et al. 2013) so that people can not afford to buy it

I. M. Raimundo (✉)
Department of Geography and Centre for Policy Analysis (CAP), Eduardo Mondlane University, Maputo, Mozambique
e-mail: ines.raimundo@uem.mz

© The Author(s), under exclusive license to Springer Nature Switzerland AG 2022   47
L. Montedoro et al. (eds.), *Territorial Development and Water-Energy-Food Nexus in the Global South*, Research for Development,
https://doi.org/10.1007/978-3-030-96538-9_4

2. The size of the plot of land that is available to access, this impulsed by the land evaluation
3. The availability of land in nearby provinces, such as Maputo and
4. With a lesser extent, the existence of new bus public transports.

In this context, this essay aims to discuss the demographic characteristics of the Province, together with the city of Maputo and some of its districts which are part of the newly designated Metropolitan area of Maputo. Nowadays, Maputo city is getting more and more connected with nearby districts of its province. Toward the north, there are the districts of Marracuene and Manhiça. Toward the south are the districts of Matola, Namaacha, and Boane. Daily, thousands of people of these districts use public or semi-public or individual means of transport, including train and buses, aiming to meet their needs, namely education, health, employement, trade, and other businesses.

Administratively Maputo city and the Province of Maputo are separated. However, they are closely linked and interdependent because Maputo city can not afford to live without the Province of Maputo and vice-versa. In such interdependency, challenges can be found and foreseen. The study performed by the Centre for Policy Analysis (CAP) of the Eduardo Mondlane University (CAP 2020) stresses that along with the fast growth of the population in the Province, particularly in the new wards that arose in Matola and the Municipality of Boane, a slowness in the supply of social infrastructure and job opportunities can be observed. Maputo is in the eye of cyclones, floods, and long periods of drought to worsen the situation (INGC 2003; República de Moçambique-Conselho de Ministros 2017). This chapter aims to analyze the demographic characteristics of the Province with some data available. The chapter is based on data from censuses, thesis, climate atlas, reports, and some inferences that result from the longstanding experience of the author of being a dweller of Maputo city. It would be interesting if the study could fully present the required information in the discipline of Demography. The required demographic information is the total number of people currently living in the Province, based on key questions such as: How many people are born? How many are dying? Furthermore, how many are migrating?

Demographers or those who study population dynamics such as Weeks (1986) suggest that comprehensive demographic analysis should be based on: (1) Population size and distribution; (2) population processes, which includes fertility, mortality, and migration; and (3) population structure and characteristics.

Ideally this could be done also for our study area, but the restricted access to demographic information limits the essay to the most recent data collected during the census. In addition, the inaccessibility of highly classified information hinders demographic justice to a study of all the provinces of Mozambique.

2 Geographic Location, Landscape, and Administrative Framework

The Province of Maputo is the most southern province of Mozambique, bordered by the Gaza Province on the north, on the east by the Indian ocean and by the City of Maputo, capital of the Republic of Mozambique. Reading the Geographic Atlas of Mozambique (1986) it is observable that Maputo is bordered by the e-Swatini Kingdom (west) and the Republic of South Africa (south). The administrative centre of the Maputo Province is the city of Matola.

The Province of Maputo is formed by eight districts, namely Boane, Magude, Manhiça, Marracuene, Matola, Matutíne, Moamba, and Namaacha. The capital is the city of Matola (Governo da Província de Maputo 2013). The Province possesses four municipalities: Matola city, Boane, Manhiça, and Namaacha villages (Imprensa Nacional-BR 2020). Dos Muchangos (1999) states that Maputo Province possesses a surface of 25.765 Km^2, representing (3.2%) of the surface of Mozambique.

3 Is the Landscape Responsible for the Sparse Population Distribution of Maputo Province?

Unlike exponents of geographic determinism, such as German geographers Ratzel and Humboldt, geographers like La Blache never believed physical conditions could determine population distribution (Frabrício and Vitte 2015; De Oliveira 2019). On the contrary, it is our contention to illustrate how physical conditions influence human aspects, including population distribution (Araújo and Raimundo 2002). Physical factors relate to the quality of soils, climate, topography (landform), hydrography, proximity, or distance from the coast and from the main urban areas. At the same time, economic and social factors such as accessibility from the city centre, the job market, the degree of infrastructural investments, the availability of schools, the access to health facilities (and the closeness with South Africa in the case of Maputo) determine the distribution of the population in a given territory. The highly concentrated population of the Province of Maputo is found in the districts located closer to Maputo, in particular in Marracuene and Manhiça in the north and Matola (city) and Boane in the south.

Studying population growth by districts is the most challenging exercise in a country such as Mozambique. Available data are disaggregated up to the level of the provinces, and hardly include the districts. However, exceptions happened to some districts based on their economic and demographic performance. For example, Matola, also a district, is classified as category "B" due to its economic and demographic dynamism. For this reason, the Government of Mozambique reclassified in 2020 the city of Matola to "B" category likewise Beira and Nampula (Conselho de Ministros Resolução 21/2020-BR 2020a, b). Meanwhile, the exact Resolution of the

Table 1 Population distribution by districts of Maputo Province between 2007 and 2017

#	District	Size (Km²)	Population in 2007	Population density (Hab/Km²)	Population in 2017	Population density (Hab/Km²)
1	Boane	820	102,457	124.94	210,367	256.54
2	Magude	6,960	54,252	7.79	62,297	8.95
3	Manhiça	2,380	157,642	66.23	205,053	86.15
4	Marracuene	666	157,642	236.69	218,788	328.51
5	Matola	373	671,556	1,800.41	1,032,196	2,767.28
6	Matutúine	5,387	37,239	6.91	43,664	8.10
7	Moamba	4,628	56,746	12.26	88,583	19.14
8	Namaacha	2,144	41,954	19.56	47,129	21.98

Source INE (2007, 2019); CAP (2020)

Council of Ministries of Mozambique upgraded the districts of Boane, Manhiça, and Marracuene to the category "B" of districts (Resolution 22/2020 BR 2020a, b).

Table 1 reveals that the lowest population is in Magude, Matutúine, Moamba, and Namaacha. Interestingly, these districts share borders with South Africa, being Namaacha the only district that shares borders with both e-Swatini and South Africa. All districts are in the driest area of the Maputo Province. The lower numbers on their population—apart from the birth rate that might be lower—is presumably related to the outcoming migration flows to South Africa and e-Swatini.

4 Population Growth of Maputo Province Between 1980 and 2017

Mozambique post-independence census occurred since 1980, when the first census was held. In the first census, Maputo province and Maputo city were administratively part of the same area, as they were split only in 1986 (Pililão 1989). In the first census in 1980, this area was made of 500,892 (DNE 1981) and increased to 1,225,489 in 2017 (INE 2019). Therefore, it is possible to perceive an increase (58.3%) making Maputo province the fastest growing population in Southern Mozambique. Nevertheless, what is still unclear is what happened between 1997 and 2007, resulting in a decline in the province population. Data shows that from 1997 to 2007, the Province of Maputo's growth population declined by 8.93%, conversely for the fastest growth between 2007 and 2017, where it registered an increase of 52.56% (Table 2).

Before this, the last census was in 1980, which meant a 17-year gap between the censuses. This gap makes it difficult for any comparison during those 17 years. At the same time, three significant events occurred in the country:

Table 2 The Maputo Province and Southern Mozambique Population Growth between 1980 and 2017

Place	1980	1997	Population growth (%)	2007	Population growth (%)	2017	Population growth (%)
Maputo Province	500,892	806,179	60.94	1,225,489	52.01	2,507,098	104.57
City of Maputo	537,912	966,837	79.73	1,111,638	18.07	1,088,000	−2.12
Gaza Province	982,603	1,062,380	9.11	1,216,284	20.22	1,422,000	20.57
Inhambane Province	1,023,879	1,123,079	9.78	1,271,818	13.24	1,496,824	22.50
Mozambique	12,130,000	15,278,300	25.95	20,226,900	32.38	27,909,798	37.98

Source DNE (1981); INE (1999, 2007, 2019)

Table 3 Population growth rate of the Maputo Province between 1980 and 2017. Calculated based on the formula $r = \sqrt{(Pt + \frac{n}{Pt}} - 1$ where (r) is equal to the rate of population growth; (n) is census interval; (Pt + 1) is population ten years later; and (Pt) population in the initial year

#	Population	Population growth rate
1980	500,892	–
1997	806,179	26.8
2007	1,225,489	23.2
2017	2,507,098	43.3

(a)   In 1986, there was a reclassification of urban areas;
(b)   The so-called "civil war" (1976–1992) and,
(c)   The return of the refugees and Internally Displaced Persons.

Table 3 shows the population growth rate. It is observable that Maputo province registered a faster growth between 2007 and 2017 as the growth almost doubled. Furthermore, a research report on the Study of Faster Population Growth and Pressure on Public Services: a Case study of Maputo indicates this trend of population faster growth.

CAP (2020), in a study commissioned by the Ministry of Economics and Finance of Mozambique, indicates that the Province of Maputo registered the more significant population growth rate (58.3%) in the whole country from 2007 to 2017. The report states that natural growth continues to be an essential variable that explains that faster growth. Interestingly, the ages between zero and four (0–4) and between 70 and 74 grew only 20%, unlike the age group between 35 and 39 which grew over 100%. In the meantime, the number of active population aged 15 and 59 years declined to 44% in 2017. In terms of absolute numbers, the report says that the unemployed people grew from 304 thousand in 2007 to 609 thousand in 2017. CAP study considered also the quality of services and other life conditions, and one of the most outstanding results regards housing stock and quality where people declared that it had improved over the previous ten years, while still lacking in terms of power supply, job opportunities, safety, availability of transport, sewage and sanitation. Education and health constitute a significant challenge in the province as with the growth of population, the existing education and health infrastructures were not enough for a larger proportion of school age population, including those who need health assistance.

5 Migration Pattern in the Maputo Province

In Mozambique, getting statistical data—even those based on oral sources—is among the most challenging tasks a researcher can face. In particular, getting migration data is not an easier task. Among the main reasons that migrant scholars have identified about why it is so difficult to get such information there are: (a) difficulties to collect migration data; (b) undeclaration of a migration status; (c) lack of a regular system of collecting data from migration. Migration data in the case of Mozambique are found either in censuses or from seldom studies that occur in the country, including Master's and PhD's dissertations. As a consequence, most migration data results from indirect calculations or measurements. Since I started to engage in migration studies, this has been the most significant barrier I have had in my career. Furthermore, data from censuses is "available" after or more than five years since their completion. For instance, the Fourth General Census of Population and Housing of Mozambique occurred in 2017, but the data is still unavailable.

Indirect measurements are shown to be very useful. Muanamoha and Raimundo (2018) used that method to analyze the migration trend in Mozambique between 1997 and 2007, which reflects migration between the second and the third general censuses of population and housing in Mozambique; as a result, they published the "Cartography of internal migration of Mozambique" based on censuses of 1997 and 2007. In their analysis of the Province of Maputo migration profile, they collected the following insights on in-bound and out-bound migration dynamics in the area.

A first consideration regards a comparison with other provinces: Maputo had in the year 1997 a positive balance (net migration) compared with other provinces, as it received more population than it lost, unlike other provinces in the same southern region (such as such as Gaza and Inhambane) and central provinces such as Tete and Zambezia and northern Mozambique the Province of Cabo Delgado.

Secondly, in the year 2007, that situation did not change as observed on the data. However, there were some changes in the Province of Nampula (northern Mozambique) which became an out-bound migration province. Meanwhile, in the same year, the Province of Maputo became a province that received more immigrants than the city of Maputo. Interestingly in the 2007 census (for the 2007 census, see CAP 2020) and even the census of 2017, the majority of inbound migrants of Maputo province were from other southern provinces and the Zambeze Province, located in the central region.;

Thirdly, the study of Muanamoha and Raimundo (2018) and the CAP report (op. cit) demonstrate that even in the census results of 2017, migration in the Maputo Province played an important role. There are two hypotheses for this rural–urban migration: (a) the city of Maputo is the hub of the economy of Mozambique, where all economic and social infrastructures are based. Central Hospital of Maputo, universities and well-equipped schools are located in Maputo. Ginisty (2020), in her book on "Urban services and spatial justice in Maputo", indicates that rural communities are still lacking social infrastructures, which are extensive in the city of Maputo, where poor dwellers live in the suburbs. Because of these inequalities, people are forced to live where jobs and other income can be gained; and (b) people follow the historical background of labour migration. Living in Maputo means—for the prospective labour migrants—becoming closer to South Africa as the Maputo Province shares common borders with South Africa and the Kingdom of e-Swatini. Another exciting finding of the studies is that women have become the most crucial segment of immigrants. In 2010, Raimundo (2010), in her doctoral thesis, demonstrated that migration was becoming more and more female as women, mainly from southern Mozambique provinces, were coming into Maputo not as attachés, but as migrants and were involved in informal economic activities. The published paper of Chikanda and Raimundo (2017) argues that the feminization trend of rural–urban migration responds to the growth of informal cross border activities, which is mainly a female activity. Women, most of them separated or divorced, or single mothers, including widows, found cross border activities as an alternative way of feeding their children, eventually their parents and other relatives. In the meantime, in the remaining regions of Mozambique migration, is still dominated by male migrants.

Fortly, from the point of view of migratory dynamics, in Mozambique, there are three groups of provinces: (1) Manica and Maputo Provinces are in-bound migration destinations; (2) the Provinces of Zambeze, Tete, Inhambane and Gaza tend to be out-bound migration areas; and (3) Cabo Delgado, Niassa, Nampula, Sofala and Maputo city that, from one period to another, changed from one migratory trend to another. In-bound internal migration—besides being female in southern Mozambique—has witnessed the presence of younger and single people, in particular male migrants. Most of the female migrants are involved in household work, studies, or street vendors in Maputo downtown with male counterparts. Women as students or in household works were mentioned by Raimundo (2010) in her doctoral thesis, meaning that the trend did not change much.

Finally, regarding literacy, most younger people are not formally educated and lack occupational experience. Obviously this situation makes their integration very difficult.

6 The Maputo Province as the Pivot of Irregular Migration Through Ponta De Ouro, Goba, Namaacha, and Ressano Garcia Borders

Migration in southern Mozambique is not a new phenomenon. Raimundo and Raimundo (2015) and Matusse (2009) pointed out that out-bound labour migration shaped the economy of southern Africa, where Mozambicans were engaged in gold mines extractive activities, plantations and other services where the pivotal countries were South Africa, Southern (current Zimbabwe) and Northern Rhodesia (current Zambia), Malawi and Tanzania.

As mentioned earlier, the Maputo Province is bordered by two countries and four borders. Goba and Namaacha are the entry and exit direct points to the e-Swatini Kingdom, while Ponta de Ouro and Ressano Garcia are the direct entry and exit border check-points to the Republic of South Africa. Djedje (2021), advances the causes of intensive migration in these borders in response to the of lack border control, signs of corruption and the dominium of cross border traders and traffickers. Meanwhile, Seda (2014) states that the Ressano Garcia border, which links South Africa and Maputo in the southwest part of the province, has been identified as a channel for clandestine movements that include human trafficking and other transnational crime. Seda (2014) and Djedje (2021) agreed that what facilitates these irregular crossings is the longstanding labour migration, the criminal network in Sub-Saharan Africa region.

Further, Seda (2017) points out that clandestine migration from Mozambique to South Africa through the Ressano Garcia border involves recruiters who use their vehicles—generally long distance ones—to transport interested crossers, mainly from provinces of Gaza and Inhambane. Gaza and Inhambane, likewise Maputo, are part of the longstanding labour migration that started about three centuries ago. Across Maputo borders, there are two recruiters "profiles" named *marehane* and challengers. The *marehane* are unemployed young men who operate at the bus terminal of Ressano Garcia, located at the border where they facilitate crossings to South Africa under the eyes of border authorities. These men charge 100–250 Zar for each crossing or more (Seda 2017: 71). According to Seda, challengers are residents of Moamba and Magude Districts, known as criminals who operate along the borders. They ambush travellers, especially those trying to cross the border illegally, rob them, rape women, and eventually kill them.

7 Conclusions

This essay aimed at presenting to readers the demographic dynamics of the Province of Maputo, located in the southern region of Mozambique. Population distribution is sparse due to a set of physical factors (climate, water, soils, and others) and economically linked to access to infrastructures, better education and health facilities, job opportunities and the possibility to cross-regional borders toward South Africa and e-Swatini. The most populated districts of the Province are located closer to the city of Maputo. The districts in such position are Matola and Boane towards the south, and Marracuene and Manhiça toward the north. On the other hand, Maputo population density is shaped by uneven population distribution and different socio-economic factors. Furthermore, the Province location is impacted by a set of natural events that influence population distribution. The most frequent are cyclones, floods, and droughts.

The Province of Maputo is the most dynamic Province as regards population changes or dynamics in southern Mozambique. Its capital, Matola city, is the leading destination of migrants given the fact that it is the most significant industrial area of Mozambique. From the second general census of population and housing done in1997 up to the fourth census in 2017, there is an significative change of population in the overall Province of Maputo.

According to the data presented, the population growth rate since 1997 was 26.8%. Meanwhile, in 2007 it was 23.2%, while in 2017 it doubled up to 43.3%. Of course, the reasons for such variations are still objects of speculation, but one of the reasons for sure is the inbound migration from other Provinces of Mozambique and the city of Maputo (which is another administrative and statistical unit). Furthermore, the

Province's geographical position and its historical background of labour migration show a demographic trend that will most probably continue with such intensification, with direct and long-term impacts on how to plan development and urban transformations in the province.

References

de Araújo MGM, Raimundo IM (2002) A Evolução do Pensamento Geográfico: UM Percursos na História do Conhecimento da Terra e das Correntes e Escolas Geográficas. Livraria Universitária, Universidade Eduardo Mondlane, Maputo

CAP—Centro de Análise de Políticas (2020) Estudo sobre o Rápido Crescimento Populacional e Pressão sobre os Serviços Públicos: O Caso de Maputo Província. Relatório. CAP-UEM, Maputo

Chikanda A, Raimundo IM (2017) Informal entrepreneurship and cross-border trade between Mozambique and South Africa. African Human Mobil Rev 3(2):943–974

De Oliveira G (2019) Vidal de La Blache e a Política: Somando às últimas décadas de releituras dos clássicos da Geografia. Paper presented at the 13th Associação Nacional de Pós-Graduação em Geografia (ENANPEGE) Conference "A Geografia Brasileira na Ciência-Mundo: produção, circulação e apropriação do conhecimento". 2–7 September 2019, São Paulo. Available via ENANPEGE. http://www.enanpege.ggf.br/2019/resources/anais/8/1562640935_ARQU IVO_Trabalhocompleto-Copia.pdf. Accessed 15 December 2021

Djedje AN (2021) Imigração irregular nas fronteiras de Ressano Garcia e de Goba, Sul de Moçambique e os Crimes Transnacionais. Dissertation, ACIPOL-Academia de Ciências Policiais, Michafutene

DNE—Direcção Nacional de Estatísticas (national Directorate of Statistics) (1981). Recenseamento Geral da População: Moçambique. DNE, Maputo

Dos Muchangos A (1999) Moçambique Paisagens e Regiões Naturais. Tipografia Globo Lda, Maputo

Fabrício DCB, Vitte AC (2015) "Princípios de Geografia humana" de Paul Vidal de La Blache. Revista Geografia e Pesquisa 9(1):76–79

Ginisty K (2020) Serviços Urbanis e Justiça Espacial em Mapuot. Regards-Croisés França-Moçambique. AFRAMO CHS—Associação Franco Moçambicana de Ciências Humanas e Sociais, Maputo

Imprensa Nacional—BR (Republic Official Gazette) (2020) Boletim da República-Publicação Oficial da República de Moçambique. Quinta-feira, 26 Março de 2020, I Série—no 59

INE—Instituto Nacional de Estatística (National Institute for Statistic) (2007) Terceiro Recenseamento da População e Habitação 2009. Resultados definitivos. Província de Maputo, Maputo

INE - Instituto Nacional de Estatística (National Institute for Statistic) (2019) Quarto Recenseamento da População e Habitação 2017. Resultados definitivos. Província de Maputo, Maputo

INE - Instituto Nacional de Estatística (National Institute for Statistic) (1999) Segundo Recenseamento Geral da População e Habitação. Resultados definitivos. Província de Maputo, Maputo

INGC—Instituto Nacional de Gestão das Calamidades et al (2003) Atlas for Disaster Preparedness and Response in the Limpopo Basin. INGC, Maputo

Kihato CW, Royston L, Raimundo JA, Raimundo IM (2013) Multiple land regimes: rethinking land governance in Maputo's Peri-urban spaces. Urban Forum 24(1):65–83. https://doi.org/10.1007/s12132-012-9163-z

Matusse R (2009) SADC from the Bantu migrations to the launch of the free trade area. Acadêmica, Maputo

Pililão F (1989) Moçambique: evolução da toponímia e da divisão territorial 1974–1987. Universidade Eduardo Mondlane, Maputo

Muanamoha R, Raimundo IM (2018) Cartografia das migrações internas de Moçambique. REMHU Rev Interdiscip Mobil Hum 26(54):31–59

Raimundo IM, Raimundo JÁ (2015) A migração moçambicana na África Austral: Povoamento e formação de famílias transnacionais. In: Arroyo M, Ariza da Cruz (eds) Território e circulação: a dinâmica contraditória. Annablume Editora, São Paulo, pp 239–270

República de Mozambique. Conselho de Ministros (2020a) Resolução do Conselho de Ministros n.º 20/2020

República de Moçambique, Conselho de Ministros (2020b) BR—Quinta-feira 26 de Março 2020, I Série no 59

República de Moçambique, Conselho de Ministros (2017) Plano Director para a Redução de Risco de Desastres 2017–2030. Maputo

República de Mozambique. Governo da Província de Maputo (2013) Perfis Distritais. Distrito de Matola. Available via the Maputo Province. https://www.pmaputo.gov.mz/por/A-Provincia/Perfis-Distritais. Accessed 10 November 2021

República de Mozambique. Ministério da Educação (1986) Atlas Geográfico Volume I. Ministério da Educação, Maputo

Seda FL (2014) Contradictory meanings of the border in Ressano Garcia community. Int J Migr Bord Stud 1(2):154–172. https://doi.org/10.1504/IJMBS.2014.066307

Seda FL (2017) Gestão de Fronteiras em Moçambique: Uma análise do impacto dos padrões internacionais de segurança para as regiões fronteiriças. Escolar Editora, Maputo

Weeks JR (1986) Population: an introduction to concepts and issues, 3rd edn. San Diego State University—Wadsworth Publishing Company, Belmont, California

Integrated Multisectoral Research Programme (PIMI). Origins, Trajectories and Horizons

Carlos T. G. Trindade, Domingos A. Macucule, and João T. Tique

Abstract The Integrated Multisectoral Research Programme (PIMI), implemented under the coordination of the Faculty of Architecture and Physical Planning (FAPF) of the Eduardo Mondlane University (UEM), is an experience of interaction between research, didactics and institutional capacity, which aims to contribute to the elaboration of a sustainable and participatory model of land analysis and planning for local territorial development. In order to address the criticalities related to Spatial Planning in Maputo Province, mainly with regard to the sustainability of the actions that take place or are carried out in this area, the PIMI started from the assumption that it is only possible to develop such a sustainable and participatory model if these research/didactics/institutional interactions allow a process of capacity-building of decision-making institutions, organizations and enterprises operating in this territory. Such process will occur, thanks to the strong participation of local technicians, through the elaboration of a Territorial Model, which will not only be the graphic expression of the spatial planning for the region under study but will also be a basis for the elaboration of future spatial planning instruments for the region.

1 Notes on the PIMI Programme

The PIMI initiative, pioneered in UEM in linking a study and research activity with two postgraduate courses of different levels, was an opportunity for a first implementation of the Ph.D. Course and the Master Course, both directed to the Territorial Planning of Regions, coordinated by the FAPF itself, which are, therefore, indispensable instruments for the success of the project that also allow training teachers of UEM and professionals.

The methodology designed for the effective operationalization of the PIMI in the field, where 'the participation, from the initial phases of the programme, of government entities and representatives of economic associations of category will

C. T. G. Trindade (✉) · D. A. Macucule · J. T. Tique
Architecture and Physical Planning, Eduardo Mondlane University, Maputo, Mozambique
e-mail: carlos.trindade@uem.ac.mz

© The Author(s), under exclusive license to Springer Nature Switzerland AG 2022　　　59
L. Montedoro et al. (eds.), *Territorial Development and Water-Energy-Food Nexus in the Global South*, Research for Development,
https://doi.org/10.1007/978-3-030-96538-9_5

be essential', took into account, according to the Terms of Reference (Annex 9 to the General Activity Plan of the UEM-Italy cooperation programme),[1] the identification of five major groups of actions, namely, (a) the definition with the Maputo Province Government of a Timetable of actions to be carried out for their initiation; (b) the organization of the launching Seminar of the Integrated Multisectoral Research Programme (PIMI), taking into account the regional and territorial development objectives set by the Government; (c) the organization of integrated planning meetings with the territory actors; (d) organization of public debates on the integrated planning results; and (e) elaboration of the final version of the Territorial Plan and public presentation of the results.

Despite the various constraints that made the operationalization and implementation of the PIMI within the timeframe planned (March 2018/March 2021) unfeasible, as well as the restrictions to prevent the spread of COVID-19, which led to the request for an extension (March 2021/March 2023), the programme is now in a crucial phase of participation by the Government entities. The programme was welcomed by the Provincial Directorates and assumed by the Governor of Maputo Province and his executive, thus allowing the start, according to the Terms of Reference of the PIMI (Annex 9 to the General Activity Plan), of '(…) integration of the UEM in the national territorial planning processes (…)'.

2   Contextualizing the PIMI

Mozambique and Italy have decided to establish, under the Agreement signed on 4 March 2011, a 'Fund for Applied and Multisectoral Research' (FIAM), as part of the 'Support Programme to the Eduardo Mondlane University (UEM) for academic reform, technological innovation and scientific research' funded by the Italian Government and integrated in the UEM Strategic Plan.

The Technical Annex to that same Agreement defines that part of the fund 'will be used to finance a multidisciplinary research finalized by the territorial development of a rural area of Mozambique, identified in consultation with the Government, which should focus in orientate the investments, public and private, necessary for the successive development phase'. This is how the Integrated Multisectoral Research Programme (PIMI) was designed, whose Terms of Reference (Annex 9 to

---

[1] In order to implement the Eduardo Mondlane University (UEM) Support Programme for academic reform, technological innovation and scientific research, financed by the Italian Government and integrated in the UEM Strategic Plan, a Cooperation Agreement between Mozambique and Italy was signed on 4 March 2011. Under this Agreement, a Fund for Applied and Multisectoral Research (FIAM) was set up to promote the quality and relevance of scientific research carried out by the UEM. The Technical Annex to the Cooperation Agreement defines that from this fund … a part will be used to finance a multidisciplinary research finalized to the territorial development of a rural area of Mozambique…. Finally, the Annex 9 to the General Activity Plan of the UEM Support Programme defines the Terms of reference for the implementation of the Multisectoral Integrated Research Programme (PIMI).

the General Activity Plan) emphasize that one of the assumptions of this cooperation programme is the 'integration of UEM in the national processes of territorial
planning, economic development and production/transfer of technologies applied
to innovation and productive diversification, which valorise the natural resources
complex of the country' (ibidem). The Terms of Reference of the PIMI also define
that the motto of the programme as the 'valorisation, preservation, sustainable use
of environmental and territorial resources for which the UEM should be able to
strengthen its analysis, research and intervention capacities, as an instrument for the
promotion of investments' being the PIMI 'the first attempt of the UEM as a whole to
propose itself as the main instrument of local sustainable development, coordinating
itself with the actors that act in various ways in the territory and proposing innovative,
non-bureaucratic, participative, pragmatic and based on scientific knowledge of the
same territory' (ibidem).

The first criterion used for the choice of the "study area" was the 'existence
of research infrastructures of UEM and other public actors, together with ease of
access' which decisively limited the area to Maputo Province. The second criterion
was 'the presence of important communication, transport, water supply and energy
infrastructures, easily accessible, in areas that, due to their geographical position and
logistical facilities, are attractive to international investments and, at the same time,
inserted in a logic of cross-border development' (ibidem), which initially limited the
area to the districts of the south-western part of Maputo Province, namely, Boane,
Moamba and Namaacha (as shown in Fig. 1). However, the first interaction with
the Government showed the need for the study to proceed taking into account the
territory of the province. Therefore, the first territorial based inventory elaborated by

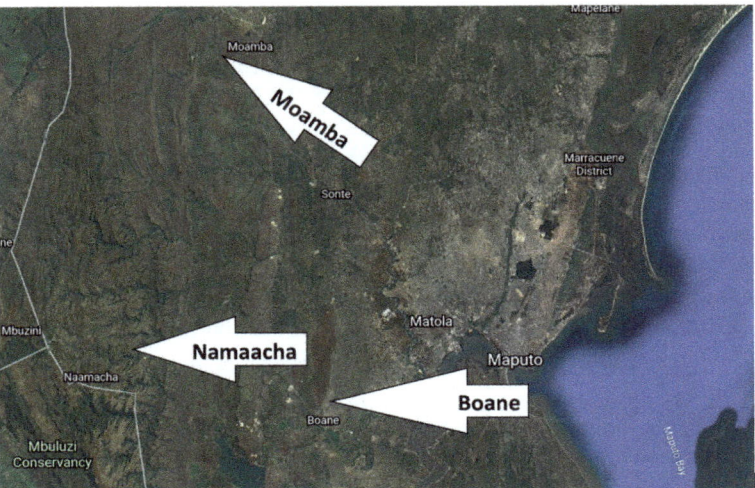

Fig. 1 The territory initially chosen for PIMI: the Boane, Namaacha and Moamba region. *Source*
elaboration by the authors on Google Earth image

FAPF under the PIMI implementation was extended to the whole Maputo Province with the objective to build a spatial database for a reading of the ongoing planning processes in the territory.

3 Research Methodology and Scope for the Implementation of PIMI

Based on the Terms of Reference (Annex 9 to the General Activity Plan), the FAPF proposed in 2016, with a view to the implementation and operationalization of the PIMI, to carry out a study called 'Study for the promotion of the integrated territorial development of the region of Boane, Moamba and Namaacha' and signed with the UEM the respective memorandum based on the project document finalized in 2018, taking into account both the collaboration of various bodies and specialties of the UEM (economics, law, social sciences, humanities, engineering, etc.) and the participation of international partners (DAStU, Politecnico di Milano).

In the project document, besides the support to the government in the elaboration and implementation of spatial planning instruments at this level, the FAPF also considered the significance to integrate in the PIMI the training of professionals and researchers to deal with the phenomenon of development of metropolitan areas and regions.

Combined with the Technical Annex to the Cooperation Agreement and the Terms of Reference (Annex 9 to the General Activity Plan), the project document prepared by the FAPF to materialize the aforementioned "Study" defines the overall objective of the PIMI as 'develop a multidisciplinary research finalized to the territorial development of a rural area of Mozambique, identified in consultation with the Government, which should undertake to channel the investments, public and private, necessary for the successive development phase'[2] (FAPF 2016, 2018) and as Specific Objectives:

1. 'to engage the different components and scientific and organisational competences in the design of a sustainable and participatory model of analysis and planning of the territory for local development, experimenting and applying new methodologies of intervention and study-action'[3];
2. 'to contribute to the integration of UEM in the national processes of territorial planning, economic development and production/transfer of technologies

---

[2] The general objective of the PIMI is defined on the basis of the Technical Annex to the Mozambique-Italy Agreement, which defines that part of the FIAM will be used for specific research activities, while a second part will be used to finance a multidisciplinary research aimed at the territorial development of a rural area of Mozambique, identified in consultation with the Government, which shall undertake to channel the public and private investments necessary for the successive development phase.

[3] The first specific objective of the PIMI, presented here, is a reformulation of the also first specific objective expressed in the project document elaborated by the FAPF.

applied to innovation and productive diversification, which enhance the natural resources complex of the country'[4];

3. 'develop a project for territorial development, together with government institutions and cooperation partners for possible funding, including funds allocated by the Government to district development'.[5]

4 The Procedures for the Operationalization of the Study Objectives

From the project document of the "Study" (FAPF 2016, 2018), for the operationalization of the PIMI and to achieve the objectives, one can enumerate the various activities closely linked to the Objectives of the "Study". Under the general objective (promotion of partnerships and investments), it is intended not only to 'conduct a multisectoral and multidisciplinary study aimed at promoting the territorial development of an area emblematic of the socio-economic dynamics that characterize the expansion areas of the largest urban centres in Mozambique, integrating knowledge about the territory at a regional scale to better understand the phenomenon and the processes involved (of socio-economic, political, cultural and environmental nature)', but also to 'elaborate an intervention model open to all possible types of partnership with the actors who interact with the territory, with the objectives of integrating and optimizing resources, and testing new working methodologies'.

In the same "Study", in the scope of study and research, seven activities are identified:

1. Develop research for the elaboration of the planning model and of a hypothesis of integrated territorial-based development plan of the area under study.

2. Implement a critical approach to the sectoral and isolated development of the territories as well as the absence of a metropolitan consciousness (Macucule 2010: 11) through the application of a research strategy based on the concepts of collaborative governance (Melatto et al. 2019: 1)[6] with a view to sustainable

---

[4] The second specific objective of the PIMI, is also the second objective expressed in the project document prepared by the FAPF.

[5] The third specific objective of the PIMI is also defined on the basis of the Technical Annex to the Mozambique-Italy Agreement signed on 4 March 2011 for the implementation of the UEM Support Programme for academic reform, technological innovation and scientific research where it is stated, regarding the PIMI that The expected result at the end of the Italian Cooperation support will be a territorial development project, to be presented to the government institutions and the community of cooperation partners for possible funding, including the funds allocated by the Government to district development.

[6] The authors, in a work aimed at systematizing studies and synthesizing existing theories on the topics of governance and smart and sustainable cities, present reflections and proposals based on the concepts of governance, collaborative governance, governance for sustainable cities and collaborative governance for smart and sustainable cities. The authors (citing Emerson et al. 2011; Healey 2015) consider the public and private sectors as the main stakeholders in collaborative governance where these actors come together to engage in consensus-driven decision-making.

development where it is assumed that 'the participation of different groups strengthens the design of sustainable and smart cities'.

3. Emphasize the socioeconomic and institutional microdynamics in the context of ongoing territorial transformations in the study area without losing sight of the structural phenomena arising from globalization, namely, migratory flows (Campos and Canavezes 2007: 96; Feijó 2017: 13), capital mobility (Sassen 2001[7] in Sassen 2005: 35), greater knowledge of information technologies (Castells 2007: 67[8] in Almeida 2009: 17; Macucule 2015: 153) and cultural (Macucule 2015: 20; HomeSpace 2012[9] in Macucule 2015: 303).

4. Develop the research considering and understanding both the notions of metropolis and the phenomenon of metropolization underway in the Maputo Region (Macucule 2010: 24). According to Ascher (1998) in (Macucule 2010: 24) the phenomenon of metropolization can be broken down into three logics: a Territorial Logic where the "spatial dynamics" of urbanization integrate the metropolization process; a Competitiveness Logic where the "economic dynamics" can, on the one hand, generate income and wealth, but on the other hand create poverty and socio-spatial problems including the degradation of the physical space; a System or Network Logic characterized by a "global dynamic" starting from the assumption that, according to Castells (2007: 520) in (Macucule 2010: 24), 'due to the nature of the new knowledge based society, organised around networks and partially formed by flows, the informational city is not a form, but a process, a process characterized by the predominance of the space of flows'.

5. Develop in-depth research on the phenomenon of metropolization occurring in the region of Boane, Moamba and Namaacha so that this event becomes known and within the agenda of land use and planning in Mozambique.

6. Contribute to make clearer the management of these territorial realities in the current political and spatial planning culture.

7. Deepen and advance the knowledge about metropolization through the understanding of hybrid territorialities that articulate the urban and the rural, the

---

[7] Sassen (2001) The Global City: New York, London, Tokyo. Princeton, New Jersey: Princeton University Press. (Originally published in 1991).

[8] Castells (2007) The Networked Society. The information age: economy, society, and culture. V. 1, 10a ed. Translation: Roneide Venancio Majer. Updated by Jussara Simões. São Paulo: Paz e Terra, 698 (Originally published in 1996).

[9] (www.homespace.dk) The research programme 'Home Space in African Cities' (2009–2012) was funded by the Danish Council for Independent Research under the management of Prof. Jorgen Eskemose Andersen of the Copenhagen School of Architecture. The programme was based on a conception and research project by Prof. Paul Jenkins of the School of the Built Environment, Heriot-Watt University/Edinburgh School of Architecture and Landscape Architecture, University of Edinburgh. It was implemented in partnership between the above institutions (led by Professors Andersen and Jenkins) and the Centre for African Studies at the University Institute of Lisbon (ISCTE)—represented by Dr. Ana Bénard da Costa—as well as the Centre for Habitat Studies and Development (CEDH) at the Faculty of Architecture and Physical Planning (FAPF) of the University Eduardo Mondlane, Mozambique—represented by Profs. Julio Carrilho, Luis Lage and Carlos Trindade.

formal and the informal, in a relational logic that emerges as a continuum that must be taken into account in the process of delimitation and management of these territorial realities.

Without neglecting to analyse the discontinuities that sometimes characterize the extensive urban field, 'making its crossing impossible and denying any possibility of continuity of meshes and development of logics of proximity' (Portas et al. 2011: 172), those 'protected environments, through which one does not pass and which we must bypass' (Mangin 2004: 330 in Portas et al. 2011: 172); and also without failing to analyse in depth the extensive urbanization phenomenon because it does not always contemplate the particularities of the rural, giving an idea of a continuum that actually does not occur, when it is more about an urbanization in the rural and not an urbanization of the rural (RUA 2002 in Ferreira et al. 2014: 495).

In turn, the objective regarding the integration of the UEM has five activities which are:

1. Academic and professional capacity building of the participants in the activities, which will have as output the activation of postgraduate courses and the training of specialists in territorial planning at a higher level;
2. Contribute to the launch of metropolitan planning in the political and scientific agenda of the country and contribute to the deepening of universal knowledge on the phenomenon of urbanization in the context of new ruralities and urbanities as a peculiar issue of the case study;
3. Contribute to the improvement and elevation of the planning and spatial planning culture in the areas and regions subjected to conurbation phenomena and to the transition to metropolitan regions, through the improvement of the legal framework, the public policies and the plans;
4. Improve the day-to-day practices of management and administration of the territory at the level of the study area;
5. Develop an integrated research plan with the didactic activities of the Master's and Doctoral programmes where the Doctoral and Master's students will be part of the Research Lines of the PIMI[10] according to the thematic in which their thesis or dissertation is inserted.

Finally, the objective related to the territorial development project is also summarized in five activities which are:

1. Integration, participation and engagement of the various actors (Stakeholders) interested and active in the territory under study in the different phases of the research, namely, Economic Associations, NGOs, Civil Society, Community Groups, etc.;

---

[10] The research is composed of five research lines, namely, Economy and Territorial Development, Spatial Planning and Management, Socio-demographic and Socio-territorial Dynamics, Governance and Public Policies and Environment and Sustainability. The Research Line is composed of the following researchers: Principal Investigator; Other Researchers; Doctoral Students; Master Students.

2. Develop a common, intermunicipal and interdisciplinary approach to planning issues taking into account that the metropolitan territory under study is continuous and there is an intense connection between the urban poles of Boane, Moamba and Namaacha, translated into great connectivity, flows or areas of influence that ignore administrative boundaries;
3. Propose urgent actions that promote integrated and sustainable development in the use of its potential in its various dimensions, creating conditions for the emergence and implementation of new economic activities and optimization of investments already made and/or to be made in order to safeguard current ecological interests and those of future generations;
4. Propose strategies that aim to promote harmonious and balanced economic and social development (Law 19/2007 of 18 July, on Spatial Planning in Mozambique);
5. Present results that can serve as an important tool for the region in the identification of new poles of development and guidance for its adequate and sustainable growth, establishing norms and criteria for occupation, clarifying the rights and expectations for development of the various sectors and actors.

5 Achieved Results

5.1 The Boa_Ma_Nhã Initiative

Within the framework of the Memorandum of Understanding between the FAPF and the Department of Architecture and Urban Studies (DAStU) of the Politecnico di Milano, the 'study for the integrated territorial development of the Boane, Moamba and Namaacha regions' was developed and finalized. The study, with funding from the "Polisocial Award 2018" aimed, among several aspects, at a contribution to the already ongoing PIMI with the aim to fill knowledge gaps and co-produce new knowledge. In June 2020 the final reports of this study were completed namely: fieldwork mission report; assessment report; planning tools report; WEF (Water-Energy-Food) case studies report; Polisocial development plan report that in its part 1 contains the territorial guidelines and scenarios articulating the WEF nexus in the Greater Maputo Region and in part 2 a local development project for Namaacha (Montedoro et al. 2020).

5.2 Assessment and Accreditation, and Commencement of Masters and Ph.D. Courses

Following the assessment of the courses by CNAQ (National Council for Quality Assessment of Higher Education) the "Declaration of Prior Accreditation" were issued on 19 December 2019 for the Master's Degree and on 16 June 2020 for the

Doctoral Degree. Both courses started on 17 August 2020 through the Zoom platform in order to avoid the spread of COVID-19, a constraint turned into a challenge. It was not foreseen that the courses would occur online, but it had to be, with use of the Zoom platform and some modules that should have been separate were carried out simultaneously with Doctoral and Master's students, enforcing the fact that the Professors were the same.

5.3 Initiation of Field Reconnaissance by Students

The first field expedition to the territory of Boane, Moamba and Namaacha, as well as Goba, organized for the students of all courses, was held on 12 December 2020. This expedition, organized with the support of the lecturers of the "Sustainability and Environment" module, had as the main objective the familiarization of the students with the study area and the observation in the field, of some of the most relevant aspects related to the concepts presented and discussed in the referred module, as well as to understand more in depth all the interactions between the physical environment and the human interventions, which manifest themselves in various ways in this territory. The expedition was organized by three main routes, namely, Maputo-Boane-Goba-Namaacha, Namaacha-Boane-Ressano Garcia and Ressano Garcia-Moamba-Maputo, expecting the students to observe during the routes aspects related to: Dynamics of Urbanization; Type/quality of buildings; Road and rail transport routes; Forms of occupation of agricultural land; Types of vegetation and vegetation cover; Relationship between Geomorphology and occupation of space; Environmental degradation (mining, erosion, buildings, coal production); The Albufeiras (Pequenos Libombos and Corumana).

5.4 The Current Status Report of The Territory of Maputo Province

The first territorial-based inventory denominated Report on the State of Spatial Planning of Maputo Province was concluded in May 2021. This inventory, which aimed to build a spatial database for a reading of the spatial planning processes underway in the territory of Maputo Province, will be discussed in the Workshop for the presentation of the study (Workshop 1) which includes multisectoral participation with different stakeholders to: (a) disseminate the purpose of the PIMI and more specifically the objectives of the research applied within the scope of the study and (b) consolidate the conceptual notes for each thematic area (namely, Territorial Economics; Governance and Spatial Planning; Territorial Development; Socio-Demographic and Socio-Territorial Dynamics; Territory and Environmental Management) having as main result a 'research guide for each thematic area'.

5.5   Interaction with the Maputo Provincial Government

The presentation of the PIMI program was welcomed by the Provincial Directorate for Territorial Development and Environment (DPDTA) at the meeting on 14 October 2020 and subsequently taken up by the Governor of Maputo Province and his executive at the Council with all the Provincial Directors held on 8 December 2020, where the FAPF also made the presentation of the PIMI programme. Quoting the Governor and his Directors, below are described the main concerns of this executive, expressed at the referred meeting, namely:

> The biggest problem in the province regards land use planning, so the initiative to train technicians in this field is commendable; there is great pressure on land for economic activities. One of the objectives of the province is to elaborate the territorial development plan. It is a unique opportunity: how to discipline the fulfilment of the designed plans. The biggest problem is even the planning of the territory, streets where not even an ambulance can pass. The idea of cooperation is very welcome; investing in human capital, further improving the capacities of officials. We want to be a model in the country as territorial development. The FAPF must use this land for Research; we are open to the PIMI . The organisation of the territory is urgent. We will consider the idea of the Provincial Plan, we will immediately schedule the public launch of the PIMI; we want to sign a memorandum with the FAPF/UEM; we want support for some specific situations such as the section Maputo-Ponta de Ouro as well as the design of the cultural center of Gwaza Muthini. We expect the transfer of knowledge to the technicians involved.

The concerns of the Provincial Executive at the meeting on 8 December clearly showed the need to urgently develop joint actions between the UEM and the Government in order to respond to the challenges arising from the spatial planning activity in the province. These concerns led to the first joint seminar held in the city of Matola (Municipality) on 8 June 2021 and organized through a partnership between FAPF and the Provincial Directorate of Territorial Development and Environment (DPDTA). The seminar on 8 June 2021 was also an opportunity to launch the PIMI in the province as well as the signing of a Memorandum of Understanding between the Government and UEM. The event was attended by the Governor of Maputo Province, His Excellency Júlio Parruque, as well as the Magnificent Rector of UEM, Orlando Quilambo and the Vice-Rector for academic area. All the District Administrators of Maputo Province were present, as well as representatives of the District Services for Planning and Infrastructures and the Municipalities of the Province (Maputo City was not included since it is another province).

6   Spatial Planning Challenges for the Maputo Province

The population should be neither "too concentrated" because "the land will collapse" nor "overly distributed" (Alexander et al. 2013: 17). The unplanned and unbridled growth of a city coupled with rapid population growth causes that population to overburden urban areas by increasingly polluting the environment, choking traffic, depleting water supplies and living in precarious housing conditions (Alexander et al. 2013: 19). Regions and cities cannot grow indefinitely because at some point the very human capacity to manage them is called into question (Haldane 1956[11] in Alexander et al. 2013: 11).

The joint seminar held in the city of Matola on 8 June, under the theme 'Spatial Planning and Climate Resilience towards Sustainable Development' discussed critical aspects of extreme importance and relevance in the context of spatial planning, aspects which it was clear to all that must be seen as "problems" whose "solutions" must be found to avoid "collapse" derived from excessive "pressure" on urban areas. This "pressure", originated by the rapid urban and population growth, results in the various problems raised and manifested during the 8 June meeting, namely, the homogenization of self-building, generally caused by the almost non-existent housing provision,[12] which results in the occupation and consumption of vast urban areas with predominantly low-density occupation models, considerably affecting the provision of support services and infrastructures for this population due to the unbearable costs; the proliferation of "condominiums" was also pointed out as an activity that occupies and consumes land with predominantly low-density occupation models, and that their promoters should "diversify the product"; the occupation of environmentally sensitive areas prone to natural disasters, with serious consequences for those who live there and high costs for the state in response and resettlement; the replacement of the practice of preparing and implementing land planning instruments by the act of demarcating plots and subdivisions, often clandestine[13] and other times eliminating areas with agricultural potential with strong impact on both productivity and the ecosystem of the region and environmental balance (Alexander et al. 2013: 19) as well as the possibility of contact with the countryside (Alexander et al. 2013: 23), leaving as an alternative the weekend outings where cars congest the roads (Alexander et al. 2013: 23); the still existing occupation and urban expansion without any prior land management activity originating urban areas devoid of "order", often connoted as "informal" occupations or settlements (Trindade, Cani et al. 2006; MICOA 2010; UN-Habitat 2008); the existence of "many" actors in the land management process.

The coordination between the various levels of decision-making was pointed out as one of the actions that can curb the disarray in land use in the territory of the province. The availability of funds for the implementation of actions within the scope of land use planning, the technical capacity building of those involved in these activities and public–private partnerships, were three conditions pointed out as indispensable

---

[11] Haldane JBS (1956). On Being the Right Size, The World of Mathematics, Vol. II, J. R. Newman, ed. New York: Simon and Schuster, pp. 962–67.

[12] It was clear from the seminar on 8 June that there are still many questions surrounding the issue of housing: Who should provide housing? How is housing provided? How is it paid for? What is the role and duty of the state?

[13] In the context of the demarcation of plots and clandestine subdivisions, the provincial governor warned of the existence at the level of the districts of various actors under the designation of topographers who establish landmarks, even georeferenced, but which are not in accordance with the rules of territorial planning.

for the possible and correct materialization of the planned activities as well as the elaborated plans. The governor stated that the planning of the territory in the Maputo Province must know how to respond to the demand for land for industry because the region is an industrial hub, as well as for housing and agricultural production. Governor Parruque also recalled that climate change is a reality and that it is necessary to invest in land use planning to mitigate the impacts of disasters and create resilience. He drew attention to the need to embark quickly on a Rectification Planning that requires a lot of courage to correct current problems and avoid future ones. The Director of the FAPF, João Tique, who also participated in this seminar, recalled that planning the territory is an important activity, fundamental and a priority and that planning the territory is also a task of governance and anticipation. Tique also said that it is necessary to provide the provinces, regions, cities, municipalities, districts, etc., with people trained to deal with spatial planning.

7 The Actions Proposed Under the PIMI Programme

According to the project document prepared by FAPF, annexed to the memorandum signed between FAPF and UEM, it is expected through the study to propose the parameters and conditions of use of natural systems[14] and areas with specific and differentiated characteristics, or with supra-provincial spatial continuities, defined by their ecological characteristics or by economic or social development parameters, or even as a result of natural disasters that require and justify planning interventions

---

[14] As also referred to in the Regulation of the Law on Territorial Planning (Decree of the Council of Ministers no 23/2008 of 1 July) A little all over the territory of Maputo province there have clearly been unbridled actions whose effects result in the destruction of natural systems, as is the case of progressive deforestation for the production of firewood and charcoal as well as for family farming, and also mining for the production of aggregates (stone and sand) for civil construction. According to Matsinhe and Soto (2011: 48) in a study by CEAGRE the best alternative to ensure a charcoal production without negative impacts on the sustainability of forests in Mozambique is the adoption of policies that limit access to forest resources, and the reduction of felling rates from the current 0.5% to 0.21%.

at regional level.[15] The aim is to define the nature and limits[16] of the interventions by the authorities,[17] local bodies and municipalities in geographical areas or within economic situations where there are, or may be, temporary or permanent mutual influences. It is intended to contribute with proposals that allow to develop, within the area subject to study, the options contained in the national programmes, in the National Plan of Territorial Development (PNDT[18] 2021) and in the sectoral plans; to translate, in spatial terms, the major objectives of sustainable economic and social development formulated in the provincial development programmes. It also meant to

---

[15] Another study by CEAGRE in partnership with Winrock International (Sitoe et al. 2016, p. 5), in the context of the preparation of the REDD + program in Mozambique with the objective of specifically contributing to the elaboration of the REDD + Strategy, identified in Mozambique 7 direct agents of deforestation and forest degradation, namely, (i) commercial agriculture, (ii) subsistence agriculture, (iii) fuelwood and charcoal, (iv) urbanization, (v) mining, (vi) logging and (vii) livestock, with subsistence agriculture (p. 4) being the agent that had the highest impact for all regions of Mozambique as well as the development corridors; the same study further concluded (p. 32) that for reducing deforestation and forest degradation (DDF) actions such as improved governance, enforcement and land use planning are essential for the implementation of direct actions such as alternatives to shifting agriculture, alternatives to biomass energy, sustained production and efficient use of biomass. REDD + (Reducing Emissions from Deforestation, forest Degradation and enhancement of carbon stocks) is a mechanism that was agreed upon in Bali by the Conference of the Parties (COP) under the United Nations Framework Convention on Climate Change (Sitoe et al. 2013).

[16] It is essential to clearly define the role and functions of the various actors who act in the management of a territory, especially vast and constantly expanding territories because, as stated by Alexander et al. (2013: 11) governing becomes increasingly difficult as the size of a region increases. Alexander et al. further refers in this context (in the same p. 11), that if the number of inhabitants exceeds a certain limit, democracy, justice and communication may be jeopardised, and a bureaucracy may be installed, which in the reality of Mozambique's rapidly growing cities results in inefficient assistance for citizens when they seek access to public services, This has been demonstrated in recent times by various phenomena, such as the long queues in public institutions, public transport, health services, banks, etc., where the already annoying response he went to have tea, which citizens often heard when approaching a counter, has gradually evolved into there is no system.

[17] Mafra and Da Silva (2004: 38) define territorial governance as a set of efficient and balanced ways of distributing functions between governmental and non-governmental bodies, both horizontally and vertically, in order to improve the impact of public policies and give the example of the decentralization of territorial governance systems carried out between 1990 and 2000 in many Western countries with reorganization of government functions from the centre, namely, deconcentration, devolution of powers, subsidiarity and budgetary decentralization.

[18] The National Territorial Development Plan (PNDT) performs several primary functions: (i) it makes explicit the strategy and the model for organising national territory; (ii) it provides the basis for the spatial coordination of sectoral policies and for the programming of major public investments with a territorial impact; and (iii) it establishes guidelines and orientations for the definition of the Spatial Planning Policy and for the preparation of other territorial plans (https://pndt.gov.mz. Accessed on 15 December 2021).

consider measures tending to attenuate inter-district development asymmetries[19, 20] and to serve as a reference framework for integrated[21] territorial planning with a view to the economic development of resident communities, through the elaboration of District Plans for Land Use, Inter-municipal and Municipal Plans for Territorial Planning. PIMI also aimed at supporting a more conscious use of the natural resources[22] of the region, promoting the rational and integrated development in the province as another strategic objective, as well as to promote the technological[23] intensification of the provincial productive base. The strategic objectives of the PIMI are also to

---

[19] These asymmetries, according to Alexander et al. (2013 p. 18) can only be mitigated with the introduction of policies that guarantee an equal division of resources and economic development throughout the region, avoiding the worsening of the imbalance between the central (urban) areas and the countryside. In the large urban areas of Maputo Province, the economic gravity force referred to by Alexander et al. (2013) of the large urban centres is remarkable because it is in the city where there is more employment, with the city being increasingly provided with basic services and infrastructure and the countryside and peri-urban areas facing various difficulties.

[20] Diniz and Croco (2006: 10–11), referring to the Keynesian perspective and Keynesianism (economic theory developed in the 30 s by John Maynard Keynes 1883–1946, published in his work general theory of employment, interest and money in 1936), as well as the different theories on the centre–periphery relationship influenced by Keynesianism, presents several examples of state intervention to overcome regional imbalances highlighting, for example, the big-push-type investments; the installation, in backward regions, of a driving industry with the aim of activating a growth pole; the targeting of regional policies towards the industrial sector; the creation of compensation mechanisms for backward regions with the implementation of a system of tax incentives to attract companies to these regions; investment in infrastructure in these regions; and, when necessary, restricting some activities in the more developed urban centres, avoiding concentration, diverting such activities to the less favoured regions.

[21] Mafra and Da Silva (2004: 18) refer that developing a region in an integrated way means, for example, planning a project with multiple purposes to obtain a much higher development than that resulting from piecemeal projects carried out in isolation. In turn Castanho et al. (2018: 2) addressing the state of the art of integrated planning cites Sachs (2004) and Ramos (2012) referring that sustainable development integrates environmental, economic, social, political and cultural issues.

[22] It is important and fundamental to ensure that the exploitation of natural resources brings visible benefits to local communities as well as contributing to the integrated development of a region. A study by the Observatory of the Rural Environment—OMR (Serra et al. 2014) concluded on a case study that there are few or no impacts on improving the lives of communities and that there is ignorance by communities of the benefits arising from the 20% of forest exploitation fees, namely Article 102 of the Regulation of the Law on Forests and Wildlife which mentions 20% of any fee for forest or wildlife exploitation is intended for the benefit of local communities in the area where the resources were extracted, under the terms of paragraph 5 of Article 35 of Law No. 10/99 of 7 July. However, Teixeira (2018: 50–55) citing several authors (Unidade de Maneio Comunitário—UMC 1998; Sitoe and Tchaúque 2007; Nhantumbo 2004; among others) mentions some examples of projects whose benefits revert to local communities in Mozambique, namely, the Tchuma Tchatu project (developed since 1995 in Mágoe District, Tete Province) where through a Ministerial Diploma the communities and the local government share the revenues from trophy licence fees, and the Chipanje Chetu project (established by the Provincial Government of Niassa, Sanga District) an attempt to replicate the experiences of Tchuma-Tchato whose activities were suspended in 2005.

[23] Teixeira (2018: 26 and 31) recalls that to speak of technology in the productivity of a region is also to speak of planning and in the management of the territory, which means new information and communications technologies (p. 31) and information systems (p. 26). In this context, it is important

ensure sustainable competitiveness[24] in the province, to promote social and territorial inclusion[25] and to consolidate the environmental protection[26] and enhancement system, which includes the areas, values and fundamental subsystems to be integrated in the ecological structure of the area of intervention of the plan.

Consistent with this framework of objectives, the activities directly linked to the doctoral and master's degree courses is expected: through the Ph.D. course, to create and deepen the autonomous capacity in carrying out and coordinating research work, with credible and scientifically based studies, and train scientists capable of giving a highly qualified contribution, in the search and definition of innovative alternatives to design organized and creative spaces in response to the needs of human activity in the field of architecture; study the symbolic and iconographic processes of popular production; organize and create qualified spaces in the wider context of the territory

---

to draw attention to the importance of Information Technology (IT) as enablers and promoters of efficiency and productivity among different companies and local industry in regions with metropolitan characteristics as referred by Costa and Garcia (2018) in a study on the metropolitan region of São Paulo (Brazil) that aimed to investigate how the externalities produced in diversified regions generate benefits for agglomerated agents. In the same study the authors also cite the work of Jacobs (1969, the economy of cities) referring that the author drew attention to the advantages of large cities and their links with the diversity of the productive structure. Always in the context of the intensification of productivity in Maputo Province, mainly in the context of agriculture, it is also worth quoting Netto (2016: 17–18) in an article on Jane Jacobs in which the author also cites the 1969 book The Economy of Cities, saying that Jacobs brings a radical hypothesis proposing the rejection of the idea that agriculture precedes the cities and also proposing that the practice of agriculture develops and intensifies from the demand of the cities that were then emerging. Therefore, being created the conditions in several areas with agricultural potential in Maputo Province, with the presence of water courses, dams, past and current experiences, etc., considerable technological investments for the intensification of production could boost the competitiveness of the region, minimizing imports.

[24] Mafra and Da Silva (2004: 29) refer that accentuating competitiveness is the main effect of globalization and that in many countries the success of globalization is associated with decentralization combined with deconcentration in decision-making. However, these authors also point out that the implementation of decentralization cannot be only through legislation but also an effective distribution of powers, competences and responsibilities between central governments, regional and cities and that new types of association between the public and private sectors are also developed.

[25] To exclude is almost always connoted with something bad and to include in general means something positive. Therefore, a territory with exclusion and socio-territorial inequalities is a territory that does not give the citizen a chance to participate in integration or in the interests of an integrated development. Analysing social exclusion/inclusion as a territorial fact the author Heidrich (2006: 2) considers social exclusion as a mechanism of loss and mentions some examples, namely, the loss of rights and social guarantees; the lack or lack in terms of education, health, housing, etc.; the exclusion of the possibility to perform work, of the conditions for the reproduction of life; and even the impossibility of the manifestation and exposure of thought in the scope of integration.

[26] One of the actions proposed by the National REDD + Strategy in Mozambique (Sitoe et al. 2013) is the establishment of protective forests in areas of fragile ecosystems and recovery of degraded ecosystems. The CEAGRE study in partnership with Winrock International (Sitoe et al. 2016: 25) in turn draws attention to the fact that although policies in general are not in favours of deforestation, practices show the opposite, and that strengthening measures for protecting conservation areas necessarily involves linking policies and practices for exploitation and use of natural resources, but this requires an effort at coordination between institutions, which means defining clear mandates from institutions and ensuring that there is capacity at district level to reduce deforestation.

and the region; through the Master course, and is intended to ensure specialized training acquired in the area of territorial planning, through a qualification that makes them able, as professionals, to integrate or coordinate, in a competent way, teams for the preparation of territorial plans and studies at local or regional level, and acquire technical capacity to fulfil, with critical, comprehensive and multidisciplinary vision, the planning activities.

Other activities will be carried out in order to achieve the expected results, namely, the elaboration of a territorial model (see table below), which will not only be the graphic expression of the spatial planning for the region under study but will also be a basis for the elaboration of future spatial planning instruments for the region; the Elaboration of a Methodological Guide for Regional Planning; the elaboration of a technical guide for the integrated management of the Municipality-District for territorial planning; the launching of the Programme's book(s); the promotion of scientific articles and communications; the promotion of the appropriation of the PIMI results by those responsible for the management of this integrated territory, and other similar ones, namely, politicians and technicians; the promotion of courses and workshops for/with managers of public entities.

FASES DE ELABORAÇÃO DO MODELO TERRITORIAL NO ÂMBITO DO PIMI
1. REALIZAÇÃO DA PESQUISA (TEÓRICA e DOCUMENTAÇÃO) para cada area temática, incluindo estudos comparativos em diferentes contextos geográficos e culturais
1.1. Economia do Território, 1.2. Governança e Ordenamento do Território; 1.3. Desenvolvimento do Território; 1.4. Dinâmicas Sócio-Demográficas e Sócio-Territoriais; 1.5. Território e Gestão Ambiental; 1.6. Organização da Base de Dados Territorial.
2. REALIZAÇÃO DO TRABALHO EMPÍRICO
2.1. Compilação e Síntese de Estudos, Projectos, Programas, Planos, Legislação - sobre a região; 2.2. Trabalho de campo; 2.3. Trabalho de gabinete; 2.4. Construção da base de dados.
3. REALIZAÇÃO DOS PRIMEIROS ESTUDOS ANALÍTICOS (DIAGNÓSTICO) com Participacao e engajamento dos Stakeholders - Identificação e definição dos Grupos Focais
3.1. Identificação e definição dos Grupos Focais; 3.2. Organização do Workshop de validação dos resultados DOS PRIMEIROS ESTUDOS ANALÍTICOS (DIAGNÓSTICO).
4. REALIZAÇÃO DAS SINTESES TEMÁTICAS (DIAGNÓSTICO) para Compreensão dos fenómenos e processo
4.1. Sistema de competitividade (produtividade -- agricola, industrial, logistica, empresarial, etc);   4.2. Sistema Ambiental; 4.3. Sistema de Povoamentos (estrutura Demográfica, rede urbana, habitacao, equipamentos); 4.4. Sistema de Infraestruturas (redes de água, energia, telecomunicações, estradas, etc); 4.5. Sistema de Mobilidade e Transportes; 4.6. Governança - Capacidade Institucional, Gestão e Planeamento Colaborativo.
Organização dos Workshops de validação dos resultados DAS SÍNTESES TEMÁTICAS (DIAGNÓSTICO) para discutir CENÁRIOS ALTERNATIVOS bem como o Cenario desejável/Visão
5. REALIZAÇÃO DE EXERCICIOS INTEGRADOS (DOS CENÁRIOS A VISAO) para Construção do quadro lógico atraves da análise estrutural para compreensao da interaccao dos fenómenos e processos de desenvolvimento territorial
5.1. Cenários; 5.2. Cenario desejável/Visão
6. EXERCICIO DE PROSPECTIVA TERRITORIAL
6.1. Construção do quadro lógico - Opcoes Estrategicas de Base Territorial; 6.2. Definição dos Elementos de Referencia para o MODELO TERRITORIAL e elaboração do MODELO TERRITORIAL.

6.2. Definição dos Elementos de Referencia para o MODELO TERRITORIAL e elaboração do MODELO TERRITORIAL
   6.2.1. **Sistema de competitividade** (produtividade -- agricola, industrial, logistica, empresarial, etc)
   6.2.2. **Sistema Ambiental**
   6.2.3. **Sistema de Povoamentos** (estrutura Demográfica, rede urbana, habitacao, equipamentos)
   6.2.4. **Sistema de Infraestruturas** (redes de água, energia, telecomunicações, estradas, etc)
   6.2.5. **Sistema de Mobilidade e Transportes**
6.3. Elementos para uma **Governança** do Território

Elaborado por: Autores, Nov. 2021 (fonte: Faculdade de Arquitectura e Planeamento Físico, 2016 e 2018)

8   Conclusions and Recommendations

The PIMI is a programme that started with the proposal in 2016 of a research project materialized in a project document in 2018, with the intention to contribute to the elaboration of a sustainable and participative model of analysis and planning of the territory, having initially as case study the districts of Boane, Moamba and Namaacha. Today, the PIMI is bringing to the forefront the problems related to Spatial Planning in the entire Maputo Province, especially regarding the sustainability of the actions that take place or are carried out in this area. The participation of the government entities since the initial phases of the programme and the first steps taken towards an awareness of integrated planning with the actors of the territory, allowed not only to interlink a study and research activity with two postgraduate courses, namely, Ph.D. and Master but also to intensify the participation of UEM in the national processes of territorial planning.

These first steps in the province with the involvement of actors responsible for the management of the territory, at various levels, have proven that it is urgent to intensify actions aimed at the enhancement, preservation, and sustainable use of environmental and territorial resources in the province, proposing innovative methodologies based either on empirical knowledge of the same territory or on scientific knowledge. Therefore, it is also urgent to clearly define the nature and limits of the interventions of the authorities of local bodies and municipalities.

Investing in integrated, sustainable and inclusive planning for the region, associated with the presence of important infrastructures of communication, transportation, water supply, energy, etc., will produce positive externalities and bring benefits and advantages due to the diversity of the productive structure created by this integration (Jacobs 1969 in Costa and Garcia 2018).

It is necessary to intensify the training of professionals and researchers to deal with the phenomenon of development of metropolitan areas and regions, and to be able to propose and implement sustainable and participatory models of analysis and spatial planning. It is necessary to provide local bodies, at different levels, with people trained to deal with spatial planning. It is necessary to promote partnerships and attract investments, favourable to integrated territorial development and not to the sectoral and isolated (Macucule 2010 p. 11).

The phenomenon of metropolization occurring in the region must become known and within the agenda of land use and planning in Mozambique. The management of these territorial realities must be clearer in the current political and spatial planning culture. It is necessary to improve and raise the planning and spatial planning culture in areas and regions subject to the metropolization phenomena, through the improvement of the legal framework, public policies and plans.

The emergence and establishment of new economic activities, as well as the identification of new centres of development and their growth, must be planned in a sustainable manner and governed by norms and occupation criteria, with a view to safeguarding the ecological interests of present and future generations.

There must be better coordination between the various levels of decision-making and local government must make better use of the existing capacities and knowledge in the country, find and allocate resources, promote partnerships, to respond to the challenges arising from the spatial planning activity in the Province. Urban areas cannot grow indefinitely and in an unplanned and unmeasured manner, in many cases with non-legal areas and urban expansion without any previous spatial planning activity, resulting in urban areas lacking "order", which are often connoted as "informal" occupations or settlements. We need to be innovative to cope with population growth that has caused this population to overburden urban areas by increasingly polluting the environment, choking traffic, depleting water reserves and living in precarious housing conditions (Alexander et al. 2013 p. 19). Solutions must be found to avoid the "collapse" derived from excessive "pressure" on urban areas.

The provision of housing, mainly of multi-family character, should be discussed not only as a responsibility of the state but also the state itself as a facilitator, because the current occupation models with predominantly low density and inefficient urban patterns result in the occupation and consumption of vast urban areas, extending cities beyond administrative boundaries (UN-Habitat 2014, 2015) and, in the end, the provision of services and infrastructures to support this population is insufficient and deficient due to the unaffordable costs. It is worth here to quote Figueiredo (2009: 25–26) citing the report The Cost of Sprawl (1974) which reports that urbanization of high-density neighbourhoods costs 21% less (after 1000hab/ha the cost increases) than medium-density urbanization and 44% less than low-density urbanization.

Another example presented by Thompson (2013: 5) referring to the case of the Halifax Regional Municipality (capital of the province of Nova Scotia, Canada) that concluded in an estimate in 2005 that the cost per household to urbanize low-density neighbourhoods is more than three times the cost to urbanize high-density neighbourhoods, forcing the Municipality to adopt a regional plan that would accommodate in urban areas 25% of urban growth instead of the current 16%. Thompson also notes that a follow-up study by the same Municipality concluded that $66 million will be saved by 2031 if this density intensification takes effect and $715 million would be saved under a 50% scenario.

It is necessary to stop the occupation of environmentally sensitive areas prone to natural disasters, with serious consequences for those who live there, and high costs for the state in the response and resettlement. Many of these areas have agricultural potential and their elimination has a strong impact on both productivity and the ecosystem of the region and environmental balance (Alexander et al. 2013: 19). Climate change is a reality, and the occupation of inappropriate areas will require considerable investment for corrective action to mitigate the impacts of disasters and create resilience.

References

Alexander C et al. (2013) Uma linguagem de padrões: a Pattern Language. Bookman, Porto Alegre

Ascher F (1998) Metapolis—Acerca do futuro da cidade. Oeiras, Celta Editora

Campos L, Canavezes S (2007) Introdução à Globalização. Instituto Bento Jesus Caraça—Departamento de Formação da CGTP-IN, Lisbon

Castanho RA, Lousada S, Camacho R, Gómez JMN, Loures L, e Cabezas J (2018) Ordenamento territorial e a sua relação com o turismo regional. Cidades [Online], 36. Available via Cidades. http://journals.openedition.org/cidades/634. Accessed 14 December 2021

Costa AR, Garcia R (2018) Aglomeração produtiva e diversificação: um enfoque sobre os serviços de tecnologia da informação. Revista Brasileira De Estudos Urbanos e Regionais 20(2):325–343

De Almeida RF (2009) A Concentração do Poder Comunicacional na Sociedade em Rede. Dissertation. Cásper Líbero Faculty, São Paulo

Diniz C, Croco M (2006) Bases teóricas e instrumentais da economia regional e urbana e sua aplicabilidade ao Brasil. In: Diniz C, Croco M (eds) Economia Regional e Urbana – Contribuições Teóricas Recentes, CEDEPLAR-UFMG, Belo Horizonte

Emerson K, Nabatchi T, Balogh S (2011, May) An integrative framework for collaborative governance. Oxford Universtiy Press on behalf of Journal or Public Administration Research and Theory, Inc.

FAPF - Faculdade de Arquitectura e Planeamento Físico—UEM (2016) Projecto de Investigação - Estudo para a Promoção do Desenvolvimento Territorial Integrado da Região de Boane, Moamba e Namaacha

FAPF - Faculdade de Arquitectura e Planeamento Físico—(2018) Descrição Detalhada do projecto - Estudo para a Promoção do Desenvolvimento Territorial Integrado da Região de Boane, Moamba e Namaacha. Universidade Eduardo Mondlane. Programa de Cooperação Itália-Moçambique. Apoio à UEM para a reforma académica, inovação tecnológica e investigação científica. Fundo para a Investigação Aplicada e Multissectorial (FIAM). Programa de Investigação Multissectorial Integrada (PIMI)

Feijó J (2017) Ruralização das cidades ou urbanização do campo? Reflexão introdutória sobre os movimentos migratórios ruralurbanos. In: Feijó J et al (eds) Movimentos Migratórios e Relações Rural-Urbanas: Estudos de Caso em Moçambique. Alcance editores, Maputo

Ferreira A, Rua J, De Mattos RC (2014) Metropolização do Espaço, Gestão Territorial e Relações Urbano-Rurais: Algumas Interações Possíveis. Geo UERJ 25(2):477–504. https://doi.org/10.12957/geouerj.2014.14408

Figueiredo MP (2009) Análise de Custos de Urbanização. Dissertation, Universidade de Trás-os-Montes e Alto Douro

Healey P (2015) Planning theory: the good city and its governance. International encyclopedia of the social & behavioral sciences, vol. 18, 2nd edn

Heidrich AL (2006) Territorialidades de inclusão e exclusão social. In: Rego N, Moll J, Aigner C (eds) Saberes e práticas na construção de sujeitos e espaços sociais. UFRGS, Porto Alegre, pp 21–44

Macucule DA (2010) Metropolização e Restruturação Urbana: O território do Grande Maputo. Dissertation, New University of Lisbon, Department of Geography and Regional Planning

Macucule DA (2015) Processo-forma urbana: reestruturação urbana e governança no Grande Maputo. Dissertation, New University of Lisbon, Department of Geography and Regional Planning

Mafra F, Da Silva JA (2004) Planeamento e Gestão do Território. SPI-Sociedade Portuguesa de Inovação, Porto

Mangin D (2004) La Ville franchisée, formes et structures de la ville contemporaine. Editions de la Villette

Matsinhe MP, Soto SJ (2011) Relatório Final - Produção de Carvão Vegetal em Moçambique: Impacto e Opções Políticas para uma Gestão Sustentável das Florestas. CEAGRE/FAEF/UEM

Melatto R, Conti D, Da Silva LF, Lofhagen JCP (2019) Governança Colaborativa: Uma Proposta para as Cidades Inteligentes e Sustentáveis. Available via Engema. http://engemausp.submissao.com.br/21/arquivos/51.pdf. Accessed 14 December 2021

MICOA - Ministério para a Coordenação da Acção Ambiental (2010) Estratégia de Intervenção nos Assentamentos Informais em Moçambique. Maputo.

Montedoro L, Buoli A., Frigerio A (2020) Towards a metropolitan vision for the Maputo province. An agenda for an integrated and sustainable territorial development in the South of Mozambique. Maggioli, Santarcangelo di Romagna

Netto VM (2016) Jane Jacobs. Revista Políticas Públicas & Cidades 4(2):9–50. https://www.aca demia.edu/31950332/Jane_Jacobs

Nhantumbo I (2004) Maneio Comunitário em Moçambique: Evolução e Desafios para o futuro, DNFFB, Maputo

Portas N, Travasso N (2011) As Transformações do Espaço Urbano: Estruturas e Fragmentos. In: Portas N, Domingues A, Cabral J (eds) (2011) Politicas Urbanas II – Transformações, regulações e projectos. (Cap. II, Fundação Calouste Gulbenkian, Lisboa, pp 161–229

Ramos M (2012) Ambiente, Educação e Interculturalidade. Revista Tempos e Espaços em Educação, no 8, Janeiro/Junho 27–39

Real Estate Research Corporation (1974) The Costs of Sprawl: executive summary, Washington, D.C.

Rua J (2002) Urbanidades e novas ruralidades no estado do Rio de Janeiro: algumas considerações teóricas. In: Marafon GJ, Ribeiro MF (eds) Estudos de geografia fluminense. Infobook, Rio de Janeiro

Sachs I (2004) Desenvolvimento: includente, sustentável, sustentado. Garamond, Rio de Janeiro

Sassen S (2005) The global city: introducing a concept. Brown J World Affairs XI(2):27–43

Serra CM, Cuna A, Amade A, Goia F (2014) O Impacto da Exploração Florestal no Desenvolvimento das Comunidades Locais nas Áreas de Exploração dos Recursos Faunísticos na Província de Nampula. Observatório do Meio Rural—OMR, Maputo

Sitoe AA, Tchaúque FJ (2007) Trends in forest ownership, forest resources tenure and institutional arrangements in Mozambique: are they contributing to better forest management and poverty reduction? A case study from Mozambique. Maputo

Sitoe AA, Guedes BS, Nhantumbo I (2013) Linha de Referência, Monitoria, Relatório e Verificação para o REDD+ em Moçambique, Relatório do País. IIED, London

Sitoe A et al. (2016) Relatório Final - Identificação e análise dos agentes e causas directas e indirectas de desmatamento e degradação florestal em Moçambique. CEAGRE/FAEF/UEM & Winrock International

Teixeira JV (2018) A Participação das Comunidades Locais na Gestão das Florestas em Moçambique: Caso dos distritos de Montepuez, Maúa, Marrupa e Majune. Dissertation. Faculdade de Ciências Sociais e Humanas da Universidade Nova de Lisboa

Thompson D (2013) Suburban sprawl: exposing hidden costs, identifying innovations. Sustainable prosperity report. Available via Institute of Smart Prosperity. https://institute.smartprosperity.ca/library/publications/suburban-sprawl-exposing-hidden-costs-identifying-innovations. Accessed 14 December 2021

Trindade C, Cani A (eds) (2006) Mozambique report: cities without slums, analysis of the situation & proposal of intervention strategies. Maputo. Center for Habitat Studies and Development (CEDH)—Eduardo Mondlane University (UEM). Available via UN-Habitat. https://mirror.unhabitat.org/downloads/docs/4399_91753_CWS%20Moz_final%20engl%20ver%20with%20figs.pdf. Accessed 14 December 2021

UEM - Universidade Eduardo Mondlane (nd) Anexo 9 ao Plano Geral de Actividades (PGA) do Programa de Apoio a UEM. Termos de referência para a realização do Programa de Investigação Multissectorial Integrada (PIMI)

UEM—Universidade Eduardo Mondlane (nd) Anexo técnico ao Acordo entre o Governo da República de Moçambique e o Governo da República Italiana para a realização do Programa de "Apoio à Universidade Eduardo Mondlane para a reforma académica, inovação tecnológica e investigação científica". Nº 3 Apoio à Investigação aplicada e multissectorial.

UN-Habitat (2008) Mozambique urban sector profile—rapid urban sector profiling for sustainability (RUSPS). UN-Habitat, Nairobi.

UN-Habitat (2014) Urban planning discussion note 1. planned city extensions. Key tools for sustainable urban development. Available via UN-Habitat. http://unhabitat.org.ph/wp-content/uploads/2016/02/discussion_note_1_-_urban_planning_for_growing_cities_key_tools_for_sus tainable_urban_development.pdf. Accessed 14 December 2021

UN-Habitat (2015) Urban planning discussion note 3. A new strategy of sustainable neighbourhood planning: five principles. Available via UN-Habitat. https://unhabitat.org/sites/default/files/dow nload-managerfiles/A%20New%20Strategy%20of%20Sustainable%20Neighbourhood%20P lanning%20Five%20principles.pdf. Accessed 14 December 2021

UMC, DNFFB, Unidade de Maneio Comunitário, Direcção Nacional de Florestas e Fauna Bravia (1998) Lições sobre o Envolvimento da Comunidade na Gestão de Projectos de Recursos Naturais em Moçambique. Maputo

Part II
Boa_Ma_Nhã, Maputo!: a "Research by Design" Project. WEF-Sensitive Territorial Assessment and Strategic Guidelines for the Namaacha, Boane and Moamba Region

Introducing the Maputo Province. A Tentative Assemblage of Planning Tools and Visions

Alessandro Frigerio and Alice Buoli

Abstract The city of Maputo and its de facto metropolitan area could be defined as an "unknown Metropolis", fragmented in terms of administrative boundaries and governance and shaped by a complex tangle of unmapped non-formal urbanization patterns, rural-to-urban migrations, as well as national and transnational flows, and local and global interests. Moreover, climate change is strongly affecting the area; threatening agricultural activities; and making water, food, and energy (WEF) security issues to be urgently considered. What is the most appropriate scale to address this complexity? To what extent the current and past planning and governance tools have been effective and adequate to address emerging socio-spatial trends and guide future territorial development? How to reframe these tools into larger territorial visions and a *long-durée* perspective? Starting from such premises and drawing on the results of "Boa_Ma_Nhã, Maputo!" research, this chapter aims first at introducing the main cultural and methodological framework of the project, and then at addressing the complex—and yet incomplete mosaic—of the available planning documents and tools in force in the Maputo Province. The purpose is to understand their potential and unexplored synergies and surface the existing inconsistencies and gaps in between them to provide an operative background for further design-oriented explorations.

1 Problem Setting: How to Study an "Unknown Metropolis"?

As mentioned in the introductory essay of this volume, this second section aims at presenting the main analytical outcomes and operative recommendations provided by a research-by-design case study focused on the Maputo Province, and on the districts

A. Frigerio · A. Buoli (✉)
Department of Architecture and Urban Studies, Politecnico Di Milano, Milan, Italy
e-mail: alice.buoli@polimi.it

A. Frigerio
e-mail: alessandro.frigerio@polimi.it

83

and municipalities of Boane, Moamba and Namaacha. Due to the rich sequence and interdependence between the contributions of this section, this chapter seeks to provide the readers some background information about the territorial framework of the research—and "set the scene" for the following contributions.

The main common ground for these essays is indeed represented by the "Boa_Ma_Nhã, Maputo!" project, a research programme funded by the Politecnico di Milano social responsibility initiative (Polisocial Award) and the Italian Cooperation Agency (AICS, Maputo) carried out between 2019 and 2020 in partnership with the Eduardo Mondlane University (Maputo, Mozambique) (UEM). Other actors directly involved in the research activities (on site and in remote) were the NGO Progetto Mondo MLAL and the Faculty of Agricultural Sciences and Technologies at the University of Milano. As previously mentioned, the project is framed into a larger international cooperation initiative supported by Eduardo Mondlane University and AICS, with the scientific support of Politecnico di Milano, and named PIMI: *Programa de Investigação Multissectorial Integrada: Estudo para a Promoção do Desenvolvimento Territorial Integrado da Região de Boane, Moamba e Namaacha.*[1] The project involved + 20 academic staff from four different departments at Politecnico di Milano (DAStU, DICA, DEIB and DENG) plus colleagues from the Faculty of Architecture and Planning (FAFP) at UEM. The project's multidisciplinary extended research team and partners involved Italian and Mozambican experts from many different knowledge fields, such as urban planning, architecture, social sciences, water management, climate studies, energy engineering, agronomy, etc. The main aim of the research was, in fact, to propose a multi- and inter-disciplinary approach to address the development of the growing peri-urban environment of Maputo in an integrated way and considering the interdependencies between internal/transnational migrations, demographic transitions, the increasing scarcity of natural resources, climate risks, natural hazards and local economic patterns (formal and informal). Particular attention was devoted to the Water-Energy-Food (WEF) Nexus (Fig. 1), considering the potential evolution of the agriculture sector; backbone economy of the region; and in relation to the whole food system and its multiple environmental, economic, social, and cultural implications. To address such complexity and benefit from the variety of methodological and epistemological approaches and expertise at play, the project was organized into four main stages:

1. The first phase (March–July 2019) mainly involved desktop-based activities with literature review, data collection and analysis from different international databases (such as WorldPop and Global Forest Watch, among many) through modelling and mapping, and remote/ad hoc interactions with local and international referents in various disciplinary fields.

2. The second stage regarded empirical and action research activities on the field (August–September 2019, plus a second shorter mission in February 2020),

---

[1] Integrated and Multisectoral Research Program: Study for the promotion of the integrated territorial development of the Boane, Moamba and Namaacha Regions.

including direct observations, photographic and territorial surveys, several meetings and interviews with experts and local actors, as well as a research seminar
co-curated with colleagues at FAPF-UEM.

3. The third stage (October–December 2019) focused on the validation and integration of the early research results with the empirical data and evidence collected
   on site.

4. The fourth and final stage (January 2020 till the end of the project) regarded a
   reconsideration of all the analytical research outcomes with the aim of producing
   a series of trans-scalar territorial cartographies, guidelines, strategic recommendations and scenarios to orient the future development of the Maputo
   Province.

Unpacking the research workflow and methodology into such a sequence of operations and methods productively impacted our understanding of a poorly known
and represented territory, about which only a few recent socio-spatial information
and georeferenced data were available or accessible. This explorative and heuristic
approach also led us to reframe the "perimeter" of our research focus on the early
stages of the research: from the original three districts/municipalities of Boane,
Moamba and Namaacha—as the main focus of the PIMI programme—to larger
territorial fields and scales outside the administrative boundaries of the province. In
fact, a "telescopic gaze" at the conditions and challenges of the Maputo Province
brought to the surface the need to study and consider in a holistic perspective a variety
of different territorial levels and dynamics, among which:

- The transnational scale of the main water basins feeding the Maputo Province—
  the Umbeluzi, Incomati and Maputo Water Basins, shared with South African
  and eSwatini—as well as the cross-border trade along the Maputo-Johannesburg
  Corridor, and of the existing and planned natural protected areas between
  Mozambique and the bordering countries.
- The national and macro-regional scale of mobility infrastructures and energy
  production and distribution networks.
- The (still unclear) administrative interplay at all the levels of government in relation to the provision of collective and public services involving different public
  and private actors and international institutions and cooperation agencies.
- The local scale of everyday socio-spatial practices.

Working in this trans-scalar and fragmented territorial setting represented a main
productive challenge for our team (both in methodological and epistemological
terms) and a test bed for the development of innovative trans-disciplinary research
tool -boxes that guided the research from the early analytical research operations till
the design of strategic and WEF-sensitive planning visions, actions, and policies for
suggesting new ways to envision food and energy production and water management.

In addition, an essential and preliminary research operation that guided the analytical and design-oriented activities was devoted to a comprehensive and "analogical"

cartographic exploration of the province, along with the study of the current planning tools and regulations orienting the future transformation of Maputo and its
metropolitan region.

However, documents retrieval or consultation took time and several meetings
with local government officials didn't prove completely successful, so that in the
timeframe of the project it was not possible to complete the full set of existing
planning tools. Accessibility to the plans, in fact, was not always easily granted by
local institutions.

The following paragraphs are devoted to framing the main socio-spatial, economic
and ecological trends and challenges at play in the study area, as well as recombining
the main plans at the district and municipal levels, which have been made available
and accessible to our research team. What emerged is a composite mosaic of different
pieces where many territorial development trajectories are coexisting, one close to
the other, and (often) ignoring what is foreseen in the neighbouring territories, thus
failing to contribute to a common and integrated territorial vision for the province.

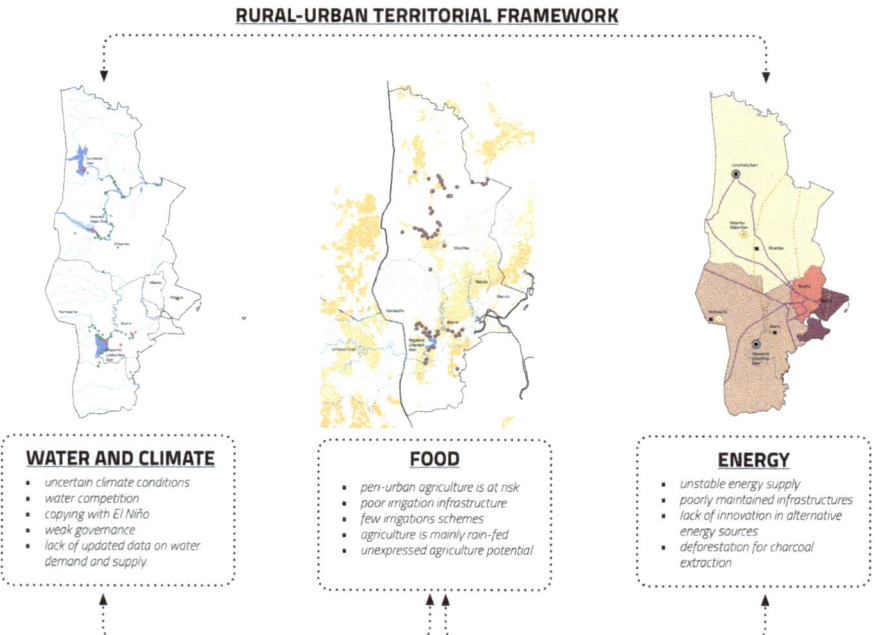

Fig. 1 Framing the Boane, Moamba and Namaacha Regions into a WEF-sensitive and rural–urban
territorial framework. *Source* Elaboration by "Boa_Ma_Nhã, Maputo!" team (A. Buoli and A.
Frigerio with D. D. Chiarelli, A. Amaranto and L. Rinaldi)

2 Territorial Planning Frameworks in Mozambique and in the Maputo Province

Maputo's growth, pushed by rural–urban migrations and conditioned by infrastructural accessibility, is following a typical centripetal development model that strongly affects the development of neighbouring districts and municipalities and produces unbalanced territorial dynamics. The districts of Boane, Moamba and Namaacha, and the municipalities of Boane and Namaacha are particularly impacted by these phenomena. However, the lack of adequate governance and institutional frameworks to face the ongoing challenges is a critical issue for the sustainable development of the province and the quality of life of its inhabitants. Existing administrative boundaries and planning tools are inconsistent in respect to the ongoing urbanization patterns, socio-economic dynamics and correspondent needs. Moreover, the governance system appears weak, with a historical lack of interest in urban development at the national government level and limited executive or facilitating capacities at the local level (Jenkins 2012, 2013).

The ongoing efforts to re-organize and implement the planning framework at all scales, both in response to the Territorial Planning Law currently in force (2007) and through alternative envisioning processes, are still lacking a trans-scalar and cross-boundary coordination able to offer an integrated territorial decision-making and planning framework. On the one hand, the aim and effectiveness of top-down planning processes, relaunched with the approval of the new law and examined in this chapter, is still extremely narrow as physical planning is seen as a technical obligation instead of a valuable potential integrated and synergic process. On the other hand, even the main large-scale strategic vision, related to the commercial relations with South Africa through the Maputo Development Corridor, is not distributing its effects, being just a fast connector to Maputo port, instead of a backbone for a networked trans-scalar system to activate synergies with local actors and communities.

In the following paragraphs, we provide an overview of the Mozambican Territorial Planning Law and of the main historical and current planning tools and conditions for the Maputo Province and the Boane, Moamba and Namaacha Districts and municipalities. For the currently in-force planning tools, in particular, we considered the Land Use District Plans (PDUT) of Boane and Moamba and the Urban Structure Plan (PEU) of the Municipalities of Maputo, Boane and Namaacha. Despite the presence of additional plans (such as the PDUT of Namaacha) not all documents have been made available to us from local authorities. However, each plan is synthetically described in its main aspects ad goals, effective or expected results and potential focuses on the Water-Food-Energy nexus, as well as the main criticalities in their implementation.

2.1 The Mozambican Territorial Planning Law

The Government of Mozambique introduced the currently in-force Territorial Planning Law (*Lei de Ordenamento do Território de Moçambique*) in 2007, followed by its Regulation with the decree 23/2008. The Law describes the use of tools and approaches to promote land use plans at national, provincial and district levels (Monteiro et al. 2017). Territorial planning is defined as the process of elaboration of plans that shape spatial forms of relations between humans and their physical and natural environment, regulating rights and forms of land use and occupation and it's considered one of the main development challenges for Mozambique for the period 2015–2035 (*ibidem*).

According to the Law and following existing literature on this topic (Pereira 2012; Sicola 2014; Monteiro et al. 2017), spatial planning in Mozambique is articulated by scale as follows:

- *National Level.* The main instrument is the National Territorial Development Plan (*Plano Nacional de Desenvolvimento*—PNDT). It defines the main priorities and guidelines for national territorial planning. The plan, under development since 2018 and approved in 2021, is focused on six drivers of development (*vectores territoriais*) and stresses the role of Maputo metropolitan system as engine of the nation (Governo de Moçambique 2019). In addition to the PNDT, there are the Special Plans for Territorial Development (*Planos Especiais de Ordenamento do Territorio*—PEOT). This instrument is meant to orient the spatial organization of areas with spatial, ecological, economic and inter-provincial continuity. Since 2007, the first two PEOT have been approved in November 2021 and are the *Plano Especial do Ordenamento do Território do Vale do Zambeze* (PEOT-VZ) and the *Plano Especial do Ordenamento do Território da Ilha de Kanyaka e de uma Parcela do Distrito de Matutuíne* (PEOT-IKPM), this latter area part of the Maputo Province. At the national level, territorial planning falls under the jurisdiction of the National Directorate of Territorial Planning from the Ministry of Land, Environment and Rural Development (MITADER).
- *Provincial Level.* The Provincial Development Plan (*Plano Provincial de Desenvolvimento Territorial*—PPDT) defines—at the provincial and inter-provincial levels—the main orientations, means and actions necessary to the territorial development. In practice, this has not been developed and used yet. At the provincial level, the PPDT is governed by the Provincial Directorate of Land, Environment and Rural Development.
- *District Level.* The District Land Use Plans (*Plano Distrital de Uso da Terra*—PDUT) are one of the main innovations of the 2007 Law and are meant to be tools that ensure territorial planning, identifying specific areas for different uses as well as to establish rules for occupancy and the use of the soil and natural resources. Yet, in many areas of the country, they are still at the initial stages of implementation (Monteiro et al. 2017). At the district level, the responsibility of the formulation of the PDUT falls under the District Services for Planning and Infrastructure (SDPI).

- *Municipal Level.* At this scale, four different tools are envisioned: the Urban Master Plan (*Plano de Estrutura Urbana*—PEU) which establishes the main norms and rules of spatial organization over the whole municipal territory considering current and potential conditions of land occupation, integrating the spatial structure at the provincial level; the General or Partial Urban Plan (*Plano Parcial/Geral de Urbanização*—PPU/PGU) which defines the general/partial organization of basic infrastructure (transport, sanitation, communication and energy) and Detailed Plan (*Plano de Pormenor*—PP) which regulates the main conditions of constructions for existing and new areas. At the municipal level, all these plans are coordinated by the local municipal authority.

Despite the existence of a hierarchical administrative architecture described and regulated by the Law, in practice, the implementation and interaction among the different planning tools and the cooperation among authorities across scales appear to be far more complicated.

For instance, looking at the case of Maputo, Maputo City is not part of Maputo Province (as better explained in the next paragraph), which in turn includes eight districts and four municipalities (Matola, Manhiça, Boane and Namaacha). The physical and political relation between districts and municipalities—when coexisting—is particularly challenging. In fact, municipalities are autonomous entities and there is a legal basis that promotes cooperation between them. In turn, the districts depend on the provincial government, the sectoral and central bodies of the state. Thus, on the one hand, horizontal relations are established between the municipalities and the state, and on the other hand, vertical relations between the districts, the provincial government and the central authorities (Fig. 2).

In this way, municipalities and districts do not have a legal and formal framework for integrating their planning tools and visions (Macucule 2016: 324). In addition, the ongoing and incomplete administrative decentralization have been hindering the horizontal integration and cooperation among the different stakeholders operating in the Maputo metropolitan region. The growing presence and importance of (non-public) actors into the spatial planning arena—such as international cooperation agencies, big transnational companies and real estate investors—have added further complexity to the already fragmented field of urban governance, often with effect of "down-grading" the role of local authorities in the decision-making process (Macucule 2016: 276–277).

In synthesis, the complexities related to the overlapping of competences and the lack of dialogue and integration of planning tools insisting on the same territories represent some of the main challenges to understand how territorial development is governed and oriented in the Maputo metropolitan area (Fig. 3).

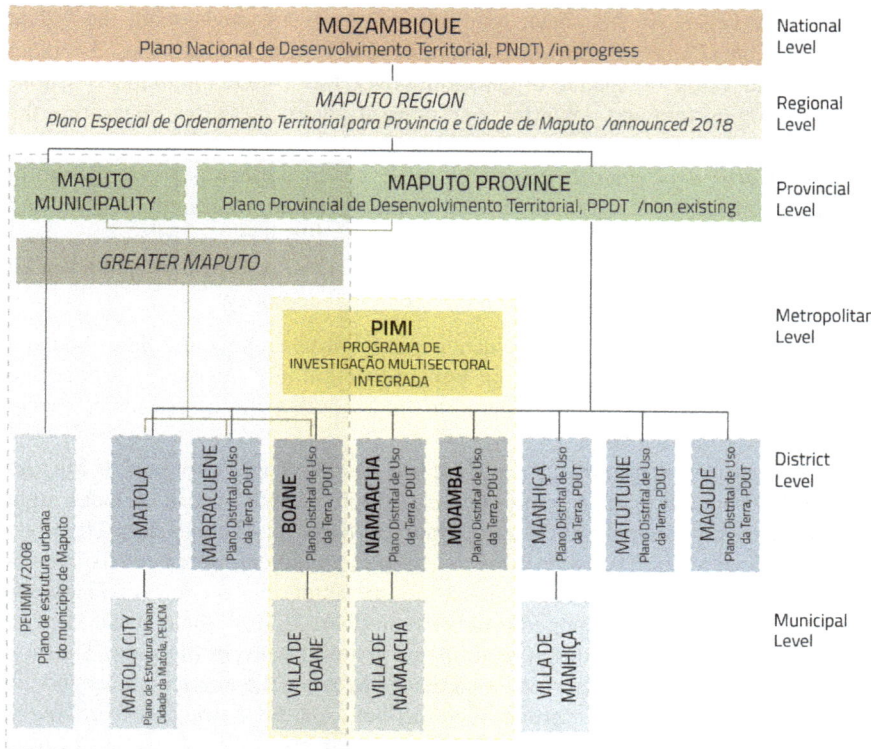

Fig. 2 Current Territorial Planning and Governance Framework in Mozambique. *Source* Elaboration by "Boa_Ma_Nhã Maputo!" team (A. Frigerio) 2020

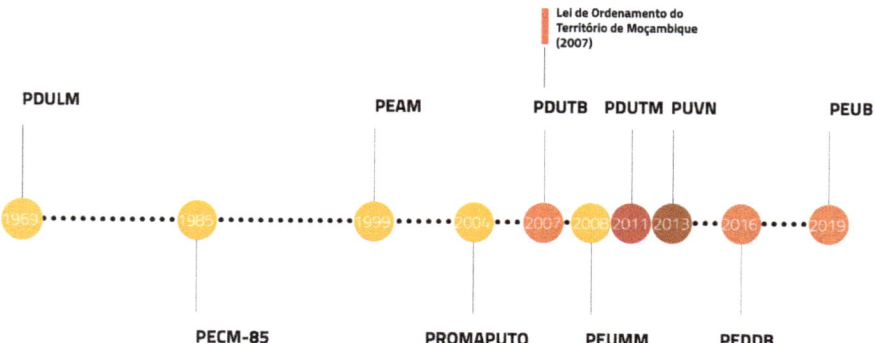

Fig. 3 Planning tools: a timeline for Maputo, Boane, Namaacha and Moamba in the last 50 years. *Source* Elaboration by "Boa_Ma_Nhã, Maputo!" team

2.2 *Maputo*

The administrative and planning history of Maputo clarifies the progressive fragmentation of its territory. Under colonial occupation, the current Maputo Province was included in the district of Inhambane. In 1908, the then province of Mozambique south of Save River was divided into two districts, Lourenço Marques and Inhambane. The district of Lourenço Marques lost much of its territory when some of its administrative divisions formed the district of Gaza in 1946. After independence, in 1974, the city was renamed Maputo and Maputo Province was formed from the district of Lourenço Marques. The city was then administratively separated by the province carrying its name in 1980. Currently, the separation between Maputo City and Maputo Province persists, despite the sprawling cross-border urbanization. The province is currently composed of eight districts: Boane, Magude, Manhiça, Marracuene, Matola, Matutíne, Moamba and Namaacha, and Matola City is its capital.

The following overview of planning tools aims at presenting the mentioned fragmentation process starting from the first comprehensive effort to plan Maputo's territory as a whole to the ongoing fragmented efforts to implement the current Territorial Planning Law.

2.2.1 Urbanization Masterplan of Lourenco Marques, 1969

This overview—which is mainly indebted with the work of Paul Jenkins, also featured in this volume, and the PIMI team—took as main starting point the Urbanization Masterplan of Lourenco Marques (*Plano Diretor de Urbanização de Lourenço Marques*)—PDULM (1969), the last colonial planning effort and the first to take into consideration the whole province as the metropolitan area of the capital city, reinforcing the city as a regional centrality (Macucule 2016: 141).

The plan laid the foundations for the urban design actions that drove the city's growth in the following decades. It foresaw a long period of validity, which could have been adapted to contemporary conditions by other strategic plans for urban restructuring with a medium-term perspective and anyway associated with the guidelines established by the PDULM. The physical planning approach is characterized by a strong modernist design, operating as on a *tabula rasa* and determining zoning based on specific functions.

From a governance perspective, the Plan also mentioned for the first time the issue of where the limits of the city and of the metropolitan region had to be located. Indeed, the area of competence of the plan was established in a tacit way, beyond the administrative boundaries existing at the time. In this sense, the PDULM inaugurated a new planning culture which suggested to operate at a "supra-municipal" level (Macucule 2016: 149).

Despite focusing on urban development, the PDULM also proposed the implementation of a series of rural and ecological protected areas in the surroundings of

Maputo. It defined a considerable amount of sensitive landscapes areas and suggested that the Inhaca Island should be oriented towards an ecological protected reserve with potential to host scientific research. The coast, on the other hand, should have been developed based on tourism and agriculture.

2.2.2 Structure Plan for the Maputo Metropolitan Area, 1999

A second important planning phase has been the failed attempt to approve the Structure Plan for the Maputo Metropolitan Area (*Plano de Estrutura da Area Metropolitana de Maputo*—PEAMM 1999) a relevant initiative in the direction of promoting a metropolitan vision for the city, beyond municipal borders and whose action was oriented towards the improvement of peri-urban areas, along with other relevant priorities.

The plan also proposed the creation of two areas mostly dedicated to agriculture and protected green lands, along with the expansion or restoration of green areas in all districts. In high-density districts, the availability of green areas should have followed the proportion of 2m^2 per inhabitant and 4m^2 in low-density areas. However, the plan was not approved due to the lack of coordination among local authorities and other main actors in the metropolitan area. One of the main problems was the absence of an entity responsible for its management and implementation. Despite these difficulties and its failed approval, the PEAMM is still a reference and important attempt in the direction of creating a coherent supra-municipal vision for Maputo.

2.2.3 Urban Structure Plan of the Municipality of Maputo, 2008

The most recent plan for the city of Maputo, coherent with the Territorial Planning Law (2007), is the Urban Structure Plan of the Municipality of Maputo (*Plano de Estrutura Urbana do Município de Maputo*—PEUMM 2008), a document produced in partnership between the Municipal Council of Maputo and the University Eduardo Mondlane, represented by the Centre for Development of Habitat Studies. The PEUMM "intends to be a guiding and coordinating instrument for the entire land occupation process in the municipal territory" (Macucule 2010: 88), with a 10-year period of action starting in 2008. The PEUMM stands out for the level of detail of analysis, diagnosis and proposition. Much of the document is dedicated to proposing new scenarios to face emerging challenges in Maputo, making considerable progress in relation to previous urban plans. The development of the PEUMM was orchestrated by a multidisciplinary network of professionals, including civil society, academia and the public sphere.

Differently from the previous plans, the PEUMM confirmed and strengthened the new administrative boundaries as confirmed by the administrative architecture ruled by the Territorial Planning Law, mentioned in the previous paragraphs.

The PEUMM recognizes the precariousness of water supply structures in the Maputo Region. As per 2008, only 75% of the Maputo population had formal access to the water network and the Umbeluzi River as main water source, responsible for serving Maputo, Matola and Boane was already close to its limit. However, the plan indicated that there was a possibility of managing underground water resources in the north of Maputo in order to remedy the water supply in some neighbourhoods. Regarding the production and distribution of energy, the plan says that the only supplier is Electricidade de Moçambique, a public company. However, the plan identified as potential and safer energy resources the construction of Mpanda Nkwada power plant and the purchase of international energy.

2.3   Boane

Boane District is located in the southwestern area of Maputo Province. It covers a surface of 806 km^2 and is crossed by the National Road n° 2 (EN2) which provides communication with Maputo and by the Salamanga railway branch (Fig. 4).

From the end of the 1990s, the Boane District has registered major projects of national impact. The main has been the construction of the MOZAL Aluminium Industry (1998), a centre of attraction for other companies that have been settled in the Beluluane Industrial Park. Direct observations on site allowed to highlight a high concentration and a sequence of activities (i.e. small and medium size commerce, medium- and large-scale agricultural production, educational facilities, residential areas and large urbanizations) along the EN2 passing through Matola Rio which forms a seamless urban continuum with Maputo, sprawling into a "rurban" environment towards Boane. This is a small size urban centre whose main axis of development appears to be the railway heading northward, as the city is limited by the Umbeluzi River in the south. An analysis of the urban growth (Montedoro et al. 2020) confirms such development trajectory, as the northern section of the city appears to have been progressively become more and more dense in terms of built environment and activities, according to both planned and unplanned urban development.

2.3.1   Boane Land Use District Plan (PDUT)

As mentioned above, the PDUT is "an instrument of district and inter-district scope, which establishes the structure of the spatial organization of the territory of one or more districts, based on the identification of areas for preferential uses and defines the norms and rules to be observed in the occupation and use of the soil and the use of natural resources" (República de Moçambique 2015: 7). The plan was developed by the District of Boane through the work of a multidisciplinary technical team composed of professionals from the National Directorate for Territorial Planning and Management (*Direcção Nacional de Planeamento e Ordenamento Territorial*—DINAPOT), the Provincial Directorate for the Coordination of Environmental Action Maputo (*Direcção Provincial da Coordenação da Acção Ambiental Maputo*—DPCAM), the District Service of Economic Activities of Boane (*Serviços*

Distritais de Actividades Económicas—SDAE) in addition to a health and education technician. The nature of the plan is mainly descriptive and analytical, devoting only the last chapter to the study of the social, economic, physical and environmental potential of the region. With a 10-year horizon, the plan remains in force and should undergo a review in 2022. The plan also includes a set of cartographies of the physical structure (water sources, soil types, agricultural infrastructure), social (education, health, religion) and urban conditions (settlements and administrative division).

One of the main potentials highlighted in the plan is the geographical location of the district. As mentioned above, Boane is cut by the EN2 that connects Maputo-Boane-Namaacha, in addition to railway lines that connects Boane to South Africa and eSwatini. In this sense, the plan envisions and supports Boane's great potential to expand its role as an economic, cultural and social corridor both on a local and international scale.

The plan also takes into consideration the geological characteristics of the district as relevant factors of economic potential, since the presence of sand sources and quarries are among the major economic highlights of the region. The PDUT identifies that Boane offers natural conditions for the expansion of the agricultural sector, having a large supply of fertile and arable land, mainly in the region of Mahanhane and 25 de Setembro.

The plan identifies, among others, the following potentials:

- Existence of very rich soils with agro-ecological aptitude ranging from high to moderate for the development of irrigated and rain-fed agriculture.
- Existence of water resources, for the supply of water for various uses to the cities of Maputo and Matola in addition to the district of Boane itself.
- Availability of areas for the practice of agriculture, the basis for the subsistence of a large part of the population.

In the field of food production and agriculture, PDUT highlights Boane as a great producer and with potential for expansion. Agriculture is the most prominent economic activity, mainly due to family production and the small-scale industrial sector. For climatic reasons, a large part of production depends on artificial irrigation, which does not yet include all producers. The PDUT identifies 14 irrigation structures in the region of 25 de Setembro, Massaca and Marien Nguabi, for example. Since 2000, some measures have been applied to enhance agricultural production, among which the construction of a forest and fruit tree nursery; the formation of 3,500 families in agro-zootechnics, credit and savings, leadership and associations; the construction of the Manguiza irrigation system, with 10 ha and supply of micro-credit inputs.

2.3.2 Boane District Development Strategic Plan (PEDD)

The 2015–2024 PEDD is a document that analyses the main social, natural and cultural aspects of the Boane District. It focuses mainly on the issue of poverty

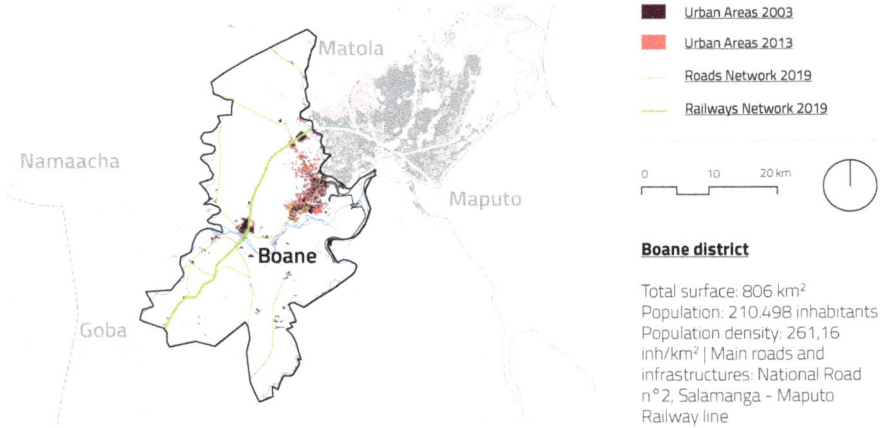

Urban Areas 2003
Urban Areas 2013
Roads Network 2019
Railways Network 2019

0   10   20 km

Boane district

Total surface: 806 km²
Population: 210.498 inhabitants
Population density: 261,16
inh/km² | Main roads and
infrastructures: National Road
n°2, Salamanga - Maputo
Railway line

Fig. 4 Urban growth in the Boane District (2003–2013). *Source* Elaboration by "Boa_Ma_Nhã, Maputo!" team (A. Buoli and A. Frigerio) based on WorldPop database (2003, 2013)

and assumes that a multi-sectoral and multidimensional strategic action is needed to improve the living conditions of the local population. The document presents a long analysis on infrastructure, sanitation, water distribution, electricity, mineral resources, agriculture and green areas. The report was organized by the local government with the support of the Regional Government of Andalusia, Spain.

"The PEDD 2015–2024, establishes a thorough and serious diagnosis of the economic and social development of the District and presents a portfolio of projects, which constitutes an instrument of action, inspiration and mobilization of all the actors involved in the battle for socio-eco- nomic and cultural heritage of the District. The Project Portfolio structures the development strategy in an integrated, compre- hensive and balanced way, seeking that planning and sectoral policies are consistent in economic and social transformation" (República de Moçambique 2016: 2).[2]

The PEDD highlights Boane's potential towards food production and water management. The availability of natural features such as water bodies, mineral resources and fertile lands is of the characteristics that makes Boane a potential reference in terms of food production and exploration. The PEDD states that Boane has 43,200 ha that could produce grains, vegetables and fruits. One of the main important points is the production of vegetables, chickens and eggs. It also states that the economic development should be aligned with social and ecological standards, specially reducing the social gap, poverty and food crises.

---

[2] Original text translated from Portuguese to English by the authors.

2.3.3 Boane Urban Structure Plan (PEU)

The PEU is primarily a document that quantifies and assesses Boane municipality's urban, geographical and social conditions (República de Moçambique, 2019). The set of instruments, composed of a written plan and a set of analytical and propositional maps, was developed by the Municipality of Boane and the participation of the studio Thomas Stellmach Planning and Architecture (TSPA, Berlin).[3] The PEU is part of a national development project (National Development Strategy 2015–2035) that aims to outline guidelines for the development of agriculture and fishing activities, expansion of industrial and extractive activities, in addition to promote sustainable development in terms of ecology, culture and historical tourism.

According to the document, the main objective of the PEU is "to support inclusive economic growth according to the national vision and economic planning, prioritizing infrastructure and spatial development that supports the goals of the PQG [Five-Year Government Plan 2015–2019] and contributes to Objective 11 of Sustainable Development" (República de Moçambique, 2019: 9).

Similarly, to the district plans for Boane, the PEU highlights Boane's privileged geographical situation as its main potential. The plan envisions Boane as an economic corridor between neighbouring countries and an informal trade centre that have been shaped to meet the needs of the large flow of people. Although the potential for economic development is high, most economic activities not related to agriculture are still of low monetary value and have little impact on local dynamics.

The agricultural potential, however, is seen as the greatest economic promise for the region. In recent years, there has been a development of agro-processing practices which, although discreet, have already shown results in agricultural production in the area. Boane has also a considerable variety of soils that could improve the production of fruits and other crops, but one of the main issues that must be addressed in order to increase and make it possible is the availability and distribution of water.

According to Thomas Stellmach (director of TSPA) interviewed by our team, one of the biggest challenges for the elaboration of the plan has been related to collecting the data and information about the current conditions in Boane, and especially in relation to climate change: "on the one hand the poorly available data on the historical hazardous events and the signals of future threats (Cyclone Idai surged during the project timeframe) forced us to find alternative solutions to understand current climate weaknesses and plan to mitigate them. Most of the information has been therefore collected with the help of the local communities through surveys and questionnaires.

---

[3] Thomas Stellmach Planning and Architecture (TSPA) is a studio based in Berlin, Germany. Emerged from Uberbau architecture & urbanism in 2014, they have an extensive portfolio of projects reflecting their experience in strategic planning, urban development and design. Their in-house team is comprised of young, talented urban planners, architects, geographers, GIS experts and urban management experts. With its main office in Berlin, TSPA operates in four world regions: CIS, MENA, Sub-Saharan Africa and Western Europe. Additionally, Thomas Stellmach is a consultant with UN-Habitat in Nairobi and cooperates closely with the urban planning lab of UN-Habitat (http://www.tspa.eu/).

Risks have been mapped and prioritized against the other planning directions" (REF. nostro report).

As regards the difficulties in implementing spatial plans, Stellmach also highlighted the importance of using hybrid methods of knowledge gathering and design (digital and empirical, remote and local), and of co-creation and co-design with local partners and authorities, as opposed to the typical western approach of imposing pre-defined design schemes. About the complexity of working in remote, Stellmach stated that "during the project development individual and team competences have been deployed complementary but separately (for proximity reasons) and the constant presence of TSPA on the site would have been beneficial especially at the project kick off, to familiarize faster with the cultural background and the organization structure of the cities" (REF. nostro report).

Finally, he underlined the criticalities of the implementation process of the plan, despite one of the main goals of the proposal was to provide the authorities with an agile and actionable tool. Stellmach recognized that "the reasons are manifolds, partially connected with the transformation of the bureaucratic structure of the entire country and partially with the limited capacity of the team to influence the local political will beyond the domain of PEU and bridge the project recommendations to the executive stage" (Boa_Ma_Nhã, Maputo, 2020b).

2.4 Namaacha

Namaacha District is located in the eastern part of the province, along the national borders between Mozambique, South Africa and eSwatini. It covers a surface of 2.156 km^2. The first modern urban settlement in the area (the city of Namaacha) dates to the colonial time (around 1850), originally developed along the EN2 connecting the city to Maputo. Thanks to the good climate, its natural amenities and the proximity to the border between the 1910s and the 1960s Namaacha became a tourist destination for Portuguese functionaries and their families. Here, a series of high-quality low-rise villas and hotels were built to respond to the demand of the European population. In addition, a series of religious orders, Catholic educational institutions and missions were established, together with the renowned sanctuary of *Nossa Senhora de Fátima* (1944) which still retains its importance as a pilgrimage destination at a regional scale, and as a landmark at the urban level (Fig. 5).

2.4.1 Urban Structure Plan for the Town of Namaacha (PEUVN)

The plan is a guiding and normative document developed in 2013 that assesses the physical, social, economic and cultural conditions of the municipality, establishing goals and priorities for urban development and social justice (República de Moçambique, 2013). This is the first initiative for planning and ordering the territory of the *Vila* after its recognition as a municipality. Due to the lack of experience in

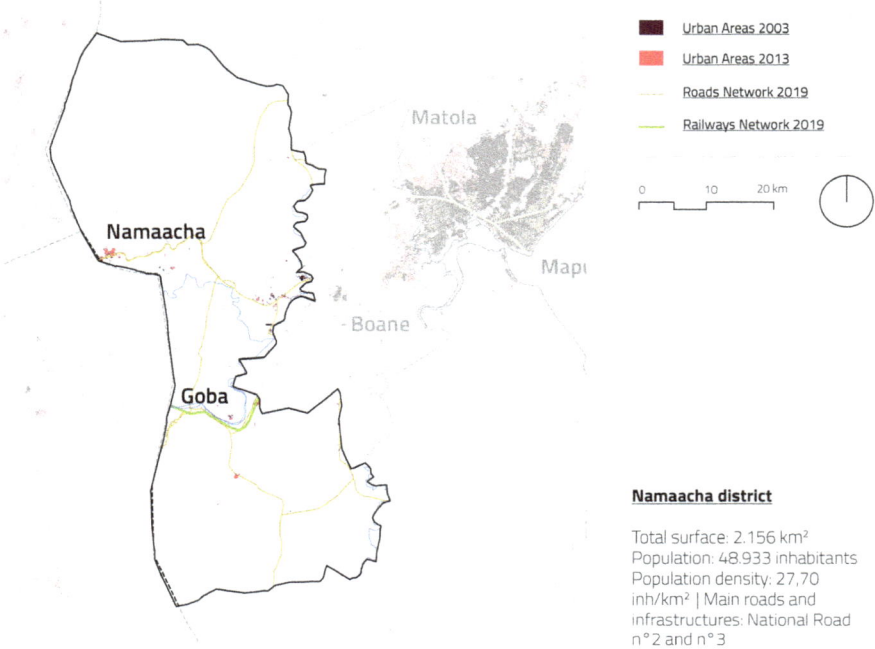

Namaacha district

Total surface: 2.156 km²
Population: 48.933 inhabitants
Population density: 27,70
inh/km² | Main roads and
infrastructures: National Road
n°2 and n°3

Fig. 5 Urban Growth in the Namaacha District (2003–2013). *Source* Elaboration by "Boa_Ma_Nhã, Maputo!" team (A. Buoli and A. Frigerio) based on WorldPop database (2003, 2013)

producing plans of this nature, the PEUVN was the result of a collaboration between various public, private and civil society actors such as the Municipal Council of Vila de Namaacha, the Habitat Development Study Centers-Faculty of Architecture and Physical Planning (Eduardo Mondlane University) and the Andalusian Agency for International Cooperation for Development.

The plan highlights the potential of Namaacha in terms of its geographical location, religious tourism and agricultural development. The location of Namaacha, on the border between Mozambique, eSwatini, and South Africa, is a characteristic that should be used to establish and intensify trans-boundary commercial exchanges between the town and the two neighbouring countries, increasing the flow of business and people. The PEUVN recognizes that the mild climate, the vegetation, the presence of waterfalls and the religious heritage can be characteristics that enhance tourism, which is still very discreet.

Regarding the environmental aspects, the PEUVN highlights the vocation and potential of Namaacha for agricultural development. The plan recognizes that in the past few decades there has been a significant loss of cultivable areas that have been converted into housing areas and in some cases simply abandoned. There is an orientation towards the development of fruit production such as strawberries and apples in periods of milder temperature and coffee, mango and orange in the warm period. Another highlight is the potential for the production of flowers by the streams. Water supply is still very precarious mainly due to the absence of permanent aquifers and stable water sources.

2.5 Moamba

Moamba District covers a surface of 4.589 km^2, and is located in the northern area of the province, 75 km northwest from Maputo and it is crossed by the EN4, connecting the city of Maputo and Ressano Garcia, the main border checkpoint with South Africa. In addition, the Maputo-Ressano Garcia railway (part of the Maputo Corridor) serves the district with daily freight and passenger trains. Together with the border town of Ressano Garcia, the city of Moamba is the main urban area of the district. Its rapid and unplanned growth has been correlated to the presence of the infrastructural network connecting Mozambique to the Republic of South Africa, turning Moamba into a corridor for people and goods. The presence of such infrastructure has been key for the urban growth of the city. Moamba's early population was constituted mainly by employees of the Railway Company and public servants from the Portuguese government along with small sections of the Mozambican population who worked for the railway company and collaborated with the colonial authorities. At the same time, direct observations on site highlighted a poor urban environment in terms of urbanity, public services and social facilities (Montedoro et al. 2020) (Fig. 6).

2.5.1 Moamba Land Use District Plan (PDUT)

The land use plan of Moamba is a guiding and normative document developed in 2011 to frame a medium- and long-term strategy for the physical development of the district (República de Moçambique, 2011). The plan has been elaborated by a technical team, composed of DINAPOT (National Directorate of Planning and Spatial Planning) technicians, DPCAM (Provincial Directorate for the Coordination of Environmental Action—Maputo) experts and a technician from the District Planning and Infrastructure Service (SDPI) of the Moamba District.

The plan presents a detailed description of the current situation and most important challenges for the district. It foresees a 10-year horizon and presents a scenario aiming at shaping the territorial development strategies for the establishment and development of infrastructure networks and equipment, defining the principles and models for the organization of the district's territory.

The plan highlights the potential of Moamba in terms of its geographical location, being the district in a privileged condition in terms of potential for projects of development for agriculture, tourism, livestock, trade and services.

In particular, the area bordering South Africa has a vocation for eco-tourism, while the area crossed by the Sabie and Incomati Rivers is suitable for agriculture and forestation projects. However, droughts, fires and deforestation are threatening

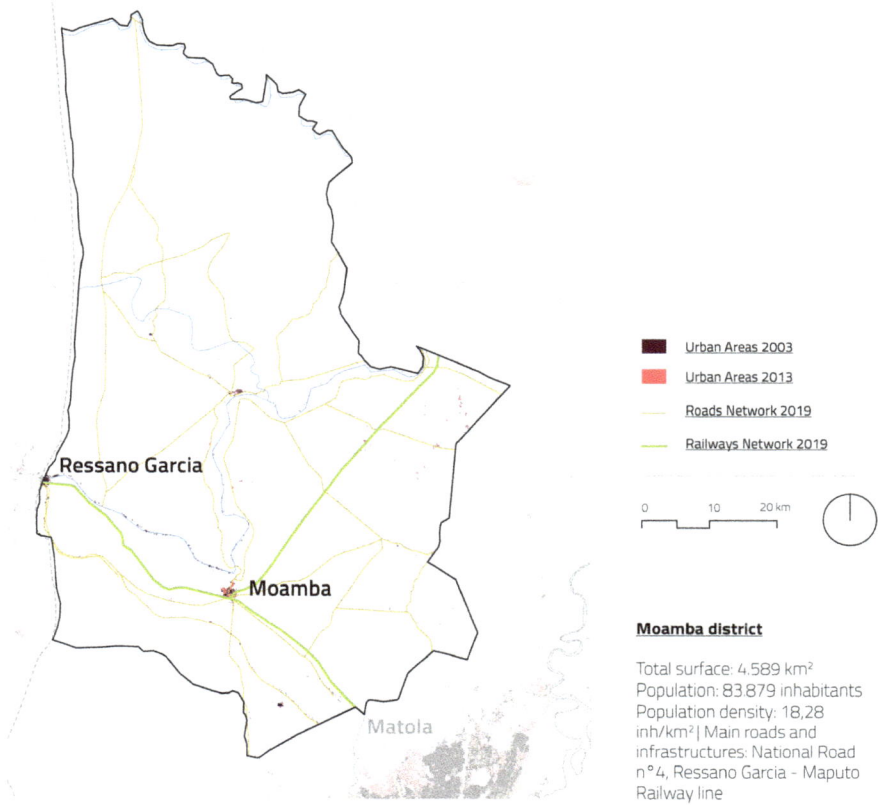

Fig. 6 Urban growth in the Moamba District (2003–2013). *Source* Elaboration by "Boa_Ma_Nhã, Maputo!" team (A. Buoli and A. Frigerio) based on WorldPop database (2003, 2013)

natural assets and a climate change adaptation strategy should be pursued, even if it's not mentioned in the plan.

The town of Moamba should invest in its potential role of trade and industrial centre along the Maputo corridor, still unexploited. The Corumana hydroelectric dam, built on the Sabie River, north of the district, supplies water for irrigation and electricity to the region and partially to the city of Maputo. It is also an important resource for fishing, contributing to local food security and providing services to the whole province. In this same perspective, it is possible to frame the project for the new large *Moamba Major* dam, under construction, that will be another crucial WEF asset for the region.

The plan also mentions technical education in agriculture and industry to support the development of local economy. A new technical school in Moamba is expected to train technical staff for the cold ice factory (crucial for vegetable conservation) and the future meat processing plant, important milestones for the metropolitan food system.

2.6 Cross-Scalar Plans and Initiatives

Out of the presented planning framework, attempts to reframe and discuss a metropolitan perspective, with the aim of overcoming the current fragmented planning condition, have been recently carried out by different kind of actors and at different scales and include the following:

- The vision promoted in the National Territorial Development Plan (2019), which introduces the concept of Grande Maputo, part of the urban subsystem of the Maputo Corridor and including Matola as a structural territorial centre, and Manhica, Namaacha and Boane as centres of rural development. Curiously, Moamba is excluded.
- The announcement in 2018 to leverage on the possibility offered by the existing law to overcome the lack of planning tools at the regional scale as an intermediate level between the national and provincial authorities, by starting the process of drafting a special planning tool for the Maputo Province and City of Maputo (*Plano Especial de Ordenamento Territorial para Provincia e Cidade de Maputo*), culminated in 2021 with the approval of the *Plano Especial do Ordenamento do Território da Ilha de Kanyaka e de uma Parcela do Distrito de Matutuíne* (PEOT-IKPM), including Inhaca Island, Machangulo Peninsula, Special Reserve of Maputo, Marine Reserve of Ponta o Oura and other areas of Matutuine district towards the South African border.
- The effort autonomously promoted by the Ministry of Transport and Communications (*Ministério dos Transportes e Comunicações*—MTC) to support the development of Maputo Corridor through a special *Programa de Desenvolvimento Espacial* (2014–15) extending to the districts of Matola, Boane, Moamba and Namaacha. The proposal also includes studies for the Lubombo Development Corridor and suggests linking the Matutuine District to create a synergic development system. This study was carried out by international consultants.
- The diffusion of the concept of Greater Maputo Area (intending the urbanized area including Maputo City but excluding Inhaca, Matola City, the eastern portion of Boane City, and the southern half of Marracuene District) to deal with specific cross-scalar metropolitan issues that emerged in some official documents and studies, such as the Comprehensive Urban Transport Master Plan for the Greater Maputo (2014) and other projects on poverty reduction and water and sanitation by different international organizations cooperating with local authorities.
- The start of specific research projects and initiatives with international organization (i.e. UN-Habitat) and cooperation agencies (i.e. AICS), to investigate strategic patterns and tools for an integrated territorial development.

The urgency and complexity of the cross-scalar challenges that Maputo has to face require that support is given through strategic and flexible government frameworks to these attempts in order to elaborate alternative planning approaches, both in terms of horizontal and vertical coordination.

3   Conclusive Remarks: Recomposing the Mosaic

Research performed in parallel by our team and the colleagues at the Eduardo Mond-lane University within the PIMI programme (see Chap. 3) allowed to analyse the existing relationships and (lack of) coordination between the different plans and initiatives to assess the state of administrative and territorial cooperation among different local authorities in the Maputo Province. From the territorial re-composition of the different plans, different readings of the spatial planning system in the Maputo Province emerged.

The early results of these operations revealed a mismatch of visions and deci-sions when analysed at the provincial level, a fact that derives either from temporal disarticulation and non-alignment of the objectives and territorial action plans in the different administrative territorial units. The analysis of the main master plans highlighted also a lack of integration between the programmatic plans (PDUT, PEU) and the intervention plans at the urban scale (PPU and PP).

The aim and effectiveness of top-down planning processes, relaunched with the approval of the Territorial Planning Law (República de Moçambique, 2007) is, in fact, still extremely narrow regarding physical planning, especially at the district level. In addition, despite the various attempts to introduce a participatory approach to the diagnostic and design stages of the planning process, this one is still seen as a technical matter, disconnected from reality, instead of a potential integrated and synergic process able to produce value at the local scale. On the other hand, even the main large-scale strategic visions—such as the ones proposed by the PDUT of Moamba and related to the commercial relations with South Africa through the Maputo Development Corridor—are not considering ways to (re)distribute the effects of such trans-scalar visions, as the Corridor still remains a trade connector bridging Johannesburg to Maputo Port, instead of being a potential backbone for a networked system able to activate synergies with local actors and communities.

As a way of conclusion (and, at the same, as an introduction to the following contributions of this section), it seems to us that the main urgent matters that should be addressed from a spatial planning and governance perspective in the Maputo metropolitan region regard the following interdependent issues, based on cultural, technical and political considerations. First, we need to build and enforce opera-tive and flexible cartographic "toolboxes" to allow planners and policy-makers to understand, represent and communicate the liveliness and the fast-changing condi-tions of the local socio-spatial dynamics at different scales. This entails, on the one hand, an issue of "cartographic justice", the development of up-to-date, open-source and accessible databases and platforms, and, on the other hand, alternative ways of depicting the urban phenomenon, which can hardly be addressed and understood (if not governed) with the current analytical and planning tools. A second issue is related to the need to build a flexible and open trans-scalar territorial agenda that could include and adapt to the fast-changing conditions of these territories. A strategic repertoire of WEF-sensitive projects and actions—within a more comprehensive general territorial "scaffolding"—could be implemented by parts according to an incremental and (financial, social and environmental) sustainability criterion, according to the potential changes in the economic, political and environmental conditions. Finally, we need to address the issue of the "implementation" of spatial plans and policies, according to a "kaleidoscopic" and "telescopic" perspective that could keep together and integrate (from the "micro-scale" of the everyday practices and lives, till the "macro-scale" of crucial environmental and societal challenges) a variety of different voices, interests and desires.

Acknowledgements The background research on the planning tools and documents presented in this chapter has been facilitated and supported by colleagues at FAPF-UEM, in particular, Dr. Elis Mavie, Prof. João Tique and Prof. Carlos Trindade, and with the additional support of Nicholas Beloso (M.A. student at Politecnico di Milano).

References

Boa_Ma_Nhã, Maputo! (2020b) Planning tools report. Politecnico di Milano, Milan
Governo de Moçambique, Ministério da Terra, Ambiente e Desenvolvimento Rural (2019) Relatórios Fase III—Proposta Técnica Premilinar de PNDT. Governo de Moçambique. https://pndt.gov.mz/index.php/relatorios-fase-iii-2/. Accessed 11 December 2021
Jenkins P (2012) Home space: context report. The Danish Council for Independent Research. http://homespace.dk/tl_files/uploads/publications/Full%20reports/HomeSpace_Context_Report.pdf. Accessed 15 November 2021
Jenkins P (2013) Urbanization, urbanism, and urbanity in an African city: home spaces and house cultures. Springer, Cham
Macucule D (2010). Metropolização e Restruturação Urbana: O território da Grande Maputo. Dissertation, New University of Lisbon, Department of Geography and Regional Planning
Macucule D (2016) Processo-forma urbana: Restruturação urbana e governança no Grande Maputo. Dissertation, New University of Lisbon, Department of Geography and Regional Planning
Montedoro L, Buoli A, Frigerio A (2020) Towards a metropolitan vision for the Maputo province. Maggioli Editore, Santarcangelo di Romagna
Monteiro J, Inguane A, Oliveira E, Joaquim S, Matlava L (2017) Territorial planning at community level in Mozambique: opportunities and challenges in a context of community land delimitation. Paper presented at the "2017 world bank conference on land and poverty". The World Bank—Washington DC, 20–24 March 2017
Pereira R P S (2012) Instrumentos de planeamento para cidades médias Moçambicanas—o caso de Pemba. Tese de Mestrado em Arquitectura—Especialização em Planeamento Urbano e Territorial. Faculdade de Arquitetura de Lisboa
República de Moçambique (2007) Lei n. 19/2007—Lei de Ordenamento do Território. Maputo
República de Moçambique. Municipio de Maputo (2008) Plano de Estrutura Urbana Do Município de Maputo.

República de Moçambique. Ministério para a Coordenação da Acção Ambiental Direcção Nacional de Planeamento e Ordenamento Territorial (2011) Plano Distrital de Uso da Terra de Moamba, vol 1 Diagnóstico da Situação Atual/vol 3 Regulamento

República de Moçambique. Município da Vila de Namaacha (2013) Plano de Estrutura Urbana da Vila da Namaacha. Diagnóstico da Situação Actual

República de Moçambique. Ministério para a Coordenação da Acção Ambiental Direcção Nacional de Planeamento e Ordenamento Territorial (2015) Plano Distrital de Uso da Terra de Boane (PDUT), vol 1 Diagnóstico da Situação Atual

República de Moçambique. Governo do Distrito de Boane (2016) Plano Estratégico de Desenvolvimento do Distrito de Boane 2015–2024

República de Moçambique. Município da Vila de Boane (2019) Plano de Estrutura Urbana. Análise da Situação Actual Drafth.

Sicola RF (2014) Ordenamento territorial e planificação estratégica no âmbito local: os sistemas de gestão do território. Vozes dos Vales 6(3):1–20. Vozes Dos Vales. http://site.ufvjm.edu.br/revistamultidisciplinar/files/2014/10/Ordenamento-territorial-e-planifica%c3%a7%c3%a3o-estrat%c3%a9gica-no-%c3%a2mbito-local-os-sistemas-de-gest%c3%a3o-do-territ%c3%b3rio.pdf. Accessed 15 December 2021

Unpacking Territorial Development in the Namaacha, Boane and Moamba Regions. A Cartographic Narrative

Alice Buoli

Abstract The chapter presents the primary analytical and methodological approach adopted and applied by the "Boa_Ma_Nhã, Maputo!" project team to represent and describe the Maputo Province in its multi-dimensional and cross-scalar features. The first part of the essay provides a brief introduction to the main debates and cultural frameworks on reading and mapping African urbanization patterns. A second section is devoted to the state of the art of the current socio-spatial and cartographic knowledge on the Boane, Moamba and Namaacha Regions. In the third section, the visions and regulations provided by the current planning plans in place in the area are combined with information from different data sources, qualitative and direct observations to propose a new cartographic knowledge base for the region. These maps have been intended as the main background to provide a more comprehensive and integrated knowledge base designed to support a revision and implementation of territorial visions and guidelines for more sustainable and WEF-sensitive development of the Maputo Province.

---

[1] Acronym of the *Centro Nacional de Cartografia e Teledetecção,* National Centre for Cartography and Remote Sensing. This is an independent institution under the Ministry of Land, Environment, and Rural Development of Mozambique. The centre is the leading and official provider of baseline GIS and satellite data and is responsible for the "direction, coordination and execution of geo-mapping and remote sensing activities at the national level, the dissemination of remote sensing techniques in the country, the acquisition, processing, and distribution of images and geo-mapping data obtained via satellite" (Source: https://www.un-spider.org/mozambique-national-cartography-and-remote-sensing-centre-cenacarta. Accessed 12 December 2021).

A. Buoli (✉)
Department of Architecture and Urban Studies, Politecnico Di Milano, Milan, Italy
e-mail: alice.buoli@polimi.it

© The Author(s), under exclusive license to Springer Nature Switzerland AG 2022
L. Montedoro et al. (eds.), *Territorial Development and Water-Energy-Food Nexus in the Global South*, Research for Development,
https://doi.org/10.1007/978-3-030-96538-9_7

1 Visualizing and Mapping African Cities: The Persistent Contentiousness of (Colonial) Boundaries

At the main entrance of the CENACARTA's[1] Headquarter in downtown Maputo, a light green and brownish pastel shaded large map of the globe welcomes the centre's visitors and users. While all the other continents appear as homogenous and border-less platforms emerging from the meridians and parallels' geographical grid, the African *plate* is crossed by its national boundaries, out of which Mozambique stands out in the far southeastern side. This small yet remarkable detail clearly visualizes the exceptional character of African political geography, providing a tangible materialization of its intrinsic *cartographic* nature.

As discussed within an extensive and rich literature,[2] African boundaries were traced by European colonial powers "in the absence of a proper knowledge of territories and people inhabiting them" Frigerio (2020). Nevertheless, they were designed as operative tools in the hands of colonial rulers to impose new spatial regimes and territorial control frameworks over autochthonous customs and practices (Scotto 2020) and, thus, enforce European interests over "topographic, demographic, and ethnographic considerations" (Amadife and Warhol 1993). Indeed, it is a shared position among African historians that colonial borders resulted from arbitrary and artificial design operations on maps, with security and land-control concerns as main drivers and criteria for tracing such lines, as opposed to existing ethnic groups composition and local political systems. These, in turn, ended up fragmented or divided by the (new) borderlines (Amadife and Warhol 1993), which have permeated ever since "the idea of colonial and post-colonial modernity and its political and planning tools throughout the twentieth century" (Frigerio 2020).

Such "cartographic creations" (Amadife and Warhol 1993) were particularly effective also in shaping post-independence nationalist movements emerging from the different colonial systems and thus confirming and reproducing the same borders, in which *long-durée* effects are still today "a hindrance to the universally sought goals of social peace and economic prosperity" (Amadife and Warhol 1993) and the primary source of inter-ethnic conflicts across the continent (Michalopoulos and Papaioannou 2011).

These macro-scale principles of (political) boundary-making were reflected at different spatial levels, in how—with the due considerations of Anglo-Saxon, French and Portuguese different spatial regimes and government systems—most African cities have been planned, built and governed throughout the twentieth and twenty-first centuries. This is the case, among many others, of the city of Maputo that—as extensively narrated in other contributions in this volume—have been designed and organized in the early twentieth century according to a principle of spatial control and ethnic segregation. The "cement city" (*cidade de cimento*) was devoted to the Portuguese lifestyle and made of permanent (concrete) buildings as opposed to the

---

[2] See for more information Scotto (2020) and Frigerio (2020).

"reeds city" (*cidade de caniço*) where the local population was supposed to live in temporary shelters (and thus subjected to eviction and demolition).

As it happened in other sub-Saharan African cities after independence, despite the establishment of a socialist government, Maputo inherited the existing colonial spatial hierarchies and structures as well as the physical and symbolic boundaries traced by the Portuguese colonists inside the city, with socio-spatial effects still visible and ruling over the organization and inhabitation patterns of the city. For instance, the limit between the cement city and the neighbourhood of Mafalala (separated by the Marien Ngouabi Avenue) still follows the original path established with the Plan of Maputo in 1903 (see, for instance, the contribution by Paul Jenkins). The road was, for a long time, not only a perceived and actual border between two different parts of the city but also a highly symbolic and political boundary synthesizing the social and spatial segregation at the basis of the city's morphogenesis. Issues of accessibility and access (to housing and basic services) are still defined and shaped by the presence of such internal urban divisions.

At the same time, other scholars (such as Mbembé and Rendall 2000) have questioned the notion of African borders as purely arbitrary colonial inventions, but instead have considered them as stratifications of more complex dynamics and socio-spatial structures including trade, religious and military corridors preceding colonization. This idea has been more recently taken on by Michel Foucher (2020), who suggested a pragmatic reconsideration of contemporary African borders as "sources of opportunity" for local economies acting as "creative interfaces, which are exploited by the trading networks that drive globalization from the bottom up". This is the case, as we will discuss in the following chapters, of the cross-boundary trade flows between South Africa and Mozambique, along the Maputo-Johannesburg corridor. Women involved in transnational trade along the corridor, called *mukheristas*, have shown themselves to be prominent economic actors, able to blend a traditional survival practice with global logistics, thus conveying a process of *glocalization* with important local impacts. Furthermore, they act as territorial agents, actively contributing to new processes of urbanization and modes of urbanity (Piscitelli 2018).

In light of these different positions and cultural trajectories, a premise is made necessary: it is not our intention—in this specific context—to contribute to (African) border theories and debates. Instead, we aim at understanding how such mechanisms of "border-making" and "border-confirming" have informed the tools and mechanisms through which—at different scales—these territories have been governed and planned and to what extent these can be considered adequate to address complex socio-spatial dynamics that have informed and are still shaping the Maputo metropolitan area.

Research conducted in the context of "Boa_Ma_Nhã, Maputo!" has indeed revealed the inadequacy of current planning tools and administrative boundaries regime.

As mentioned in the previous chapter, the new Mozambique Territorial Planning Law (2007) has reinforced the former colonial divisions, adding even more administrative delimitations (such as in the case of Maputo and Matola Cities, which are in fact indistinguishable parts of the same urban agglomeration). This can be seen as a "side effect" of the diffusion and adoption of Anglo-Saxon-inspired strategic spatial planning and (good) governance models—also pushed forward by international cooperation agencies and multi-lateral institutions such as the World Bank and UN-Habitat (Watson 2009)—which introduced even more fragmentation among the existing territorial units.

Thus, the mismatch between (a) current administrative delimitations (at different scales) inherited by the colonial government model, persisting or reinforced by current planning rules and (Anglo-Saxon-rooted) urban planning "cultures" on the one hand and (b) the pre-colonial customary rules and rights (Frigerio 2020), de facto and spontaneous urbanization and cross-boundary mobility practices, on the other hand, appears one of the most challenging issues as regards urban planning and governance in sub-Saharan urban areas. As suggested by the editors of this volume in other contexts (Frigerio 2020; Montedoro et al. 2020), the complex shaping of contemporary territorial and urban transformations in urban Africa needs to be better investigated and understood through analytical and multi-dimensional tools and media that can provide more appropriate and operative frameworks and toolboxes for envisioning more sustainable futures.

To this aim, in this chapter, we discuss the current conditions of territorial knowledge considering the (available) planning tools in the Maputo Province, an incomplete mosaic of previsions, tools and regulations (Boa_Ma_Nhã, Maputo! 2020b). This operation of interpretative assemblage allowed us to design a series of cartographic representations making evident, on the one hand, the need to reconsider such information according to larger frameworks—the transnational, the regional, the metropolitan and the *glocal*—and then the necessity to visualize the inconsistencies and contradictions between the planned land-use transformations across the different planning documents (at the district and municipal levels), in which the safeguard of natural, rural and eco-systemic resources and urban development pathways often clash between each other conspicuously.

2 Methodological Insights: Dealing with a Multiplicity of Data and Sources

According to recent research published in the context of the "Africapolis" project and database (Heinrigs 2020; OECD 2020), the African continent is "already far more urban, and its agglomerations far more numerous than the international statistics tell us". The "Africapolis" project—by using a combination of different mapping technologies and techniques interpolated with national statics and demographic data—has indeed shown that the actual number, distribution and geography of metropolitan areas and urban conurbations in the continent suggest a quite different picture from classic understandings of the urban phenomenon. This complex picture is mainly related, from one side, to the lack of a common and shared definition of what a "city" or an "urban agglomeration"[3] is in the African context, according to different national and international standards and regulations, and from the other hand, to the recurring discrepancy between administrative boundaries, urbanization dynamics and the representation of such processes, suggesting that "by simply moving the administrative boundaries of the container, it is possible to radically change the statistical representation of the contents" (OECD 2020). In addition, the authors of the study recognize the criticalities related to how competencies on collecting, analysing and producing cartographic data and information about urban dynamics: these data are often shared or co-managed by different departments, ministries or institutions, which are not always available in disseminating such knowledge with other actors.

Looking at the case of the Maputo metropolitan region, the so-called Greater Maputo—a still unofficial territorial entity which includes Maputo City, Matola City, the eastern portion of Boane City and the southern half of Marracuene District—it is possible to highlight the complexity of adopting an administrative criterion to define and represent the urban phenomenon in a fast-growing urban agglomeration. According to the official national demographic statistics, the population of Maputo city is currently slightly exceeding 1 million inhabitants while Matola has 1.6 million inhabitants (INE 2018): altogether, they form a continuous urban agglomeration of more than 2.6 million inhabitants, representing 30% of the country's urban population and the largest Mozambican urban area (de Araújo 2006). However, despite these commonalities, these territories have been organized into different administrative entities since 1988 and planned according to different tools, as we will see in the following paragraphs.

On the other hand, the approach adopted by "Africapolis" suggests other criteria to detect, identify and describe African urbanization patterns, based on a spatial analytical approach using two different sources: population data available nationally and/or locally, and satellite images collected from Google Earth (Heinrigs 2020). Similarly, "Boa_Ma_Nhã, Maputo!" research team adopted an analytical and interpretative approach that combines different spatial qualitative-quantitative research methods to fill in the territorial knowledge gaps in a context where digital cartographic and geo-referenced information is often not available or very difficult to access.

In this chapter, we will first provide some results emerging from the primary research operations performed to define an analytical multi-media and transdisciplinary mapping apparatus to represent the current conditions, trends and threats defining the Maputo metropolitan territory. We will then present some key results of such research operations that reconstruct the Maputo metropolitan area's transcalar, cross-thematic and inter-administrative images. These were later adopted as the main background for designing a series of scenarios for a more integrated WEF-sensitive spatial planning.

---

[3] Africapolis defines a continuously built-up area as "an area with less than 200 m between buildings and constructions. (...) The minimum population threshold used by Africapolis to consider an agglomeration as urban is 10 000 inhabitants. The agglomeration takes the name of the local administrative unit that is highest in the administrative hierarchy and/or population" (Source: https://africapolis.org/fr/about/defining-urban. Accessed on 12 December 2021).

3   Emerging Issues from the Cartographic Analysis: Land Competition and the Need for an Integrated Planning Framework and Cooperation

As mentioned, an essential research operation performed throughout the project was the spatialization and representation of the data about the main territorial conditions and processes occurring at different scales (demographic trends and urban growth, land cover and land uses, food production patterns, climate data, hydrographic systems, energy networks and resources, etc.) with synthetic cartographies.

The team worked mainly within a GIS environment, combining information from a variety of international institutional or independent databases (among many Global Forest Watch[4] and WorldPop[5]) and national authorities, such as the above-mentioned CENACARTA and the Eduardo Mondlane University. The research team implemented these early cartographic representations with additional data from research collected both in remote and during the fieldwork missions performed in 2019 and 2020 (Boa_Ma_Nhã, Maputo! 2020a). Gaps in the cartographic information and additional spatial analysis were integrated and implemented thanks to the processing of different plans and strategic documents provided by local administrations and the analysis and redrawing of satellite images (Cfr. Chap. 4).

Along with the juxtaposition and interpolation of different information and sources, a work of re-composition of the current planning tools in force in the study area was performed, with the aim of understanding the mutual interactions and gaps in between these plans. This operation allowed the team to visualize and investigate potential synergies, conflicts and emerging issues.

Research performed in parallel by our team and the colleagues at the Eduardo Mondlane university allowed to analyse the existing relationships and (lack of) coordination between the different plans to assess the state of administrative and territorial cooperation among different local authorities in the Maputo Province. From the territorial re-composition of the different plans emerged different readings of the spatial planning system in the Maputo Province.

The insights emerged from this interpretative work provided the starting point to locate such visions in the general assessment of the conditions and trends occurring in the area, with reference to the urban–rural relations. A series of cartographies was then produced to highlight such elements and produce operative representations allowing the team to propose alternative scenarios to the current tendencies (Cfr. Chap. 7). These cartographies are meant to construct not only a thematic overview, but also a narrative sequence that aims to allow the readers to recognize the most crucial ongoing territorial issues and emerging socio-spatial dynamics in the region.

---

[4] https://www.globalforestwatch.org/. Accessed on 12 December 2021.
[5] https://www.worldpop.org/. Accessed on 12 December 2021.

3.1 Demography, Population Density and Urban Growth

Despite the low density and mainly rural character of Mozambican demographics, in 2007–2017, the Maputo Province has witnessed outstanding growth in terms of population and urbanization (see for more details the contribution by Inês Raimundo). Indeed, according to the preliminary results of the 2017 Census (INE 2018), the overall population of the Maputo Province has doubled and is now over 3.6 million inhabitants and it is likely to double again by 2025 (Jenkins 2012). Such tendency has been the result of both high birth rates and internal migrations among Mozambican Provinces. At the same time, Maputo City has seen a decrease in the number of inhabitants of more than -3%, whose causes still need to be disclosed and discussed. Along with varied demographic trends in the different districts of the province, the population density is not homogeneous within the region: indeed, it varies from 4.000 inhabitants/km2 (Matola) to less than 30 inhabitants/km2 (Namaacha).

The first map (Fig. 1) shows the variation in demographic growth and population distribution within the Maputo Province and its districts. This is accompanied by differentiated urban growth patterns, morphologies and land occupation, mirroring different social dynamics and structures as well as different conditions of access to water, food, basic services and infrastructures (Jenkins 2012).

Available geo-referenced data from the WorldPop database (2003 and 2013) show an increase of the urbanized areas along the northern axis between Maputo and

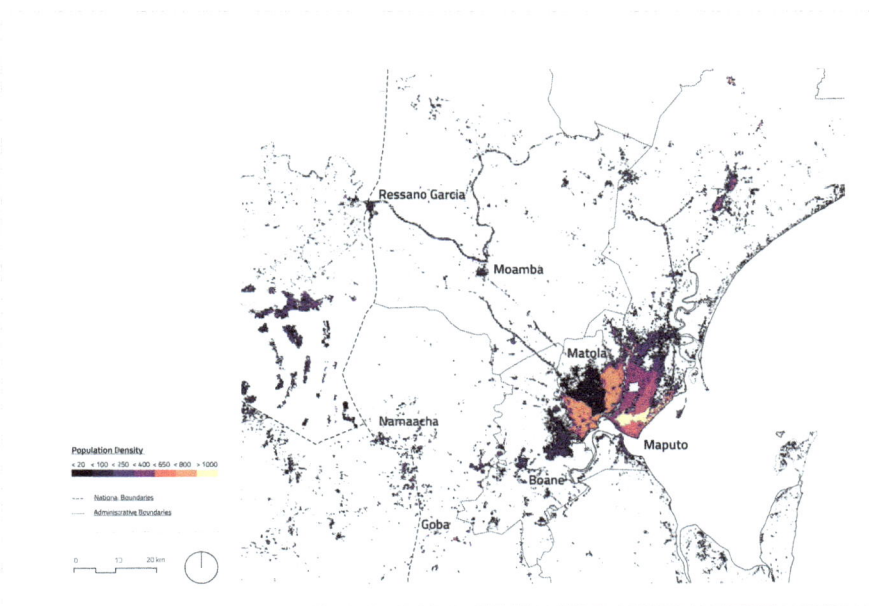

Fig. 1 Population density Elaboration by "Boa_Ma_Nhã, Maputo!" team 2019 (A. Buoli, A. Frigerio)

Marracuene, towards Boane and along the main infrastructures and rivers in the province towards the in-lands and the borders with South Africa and eSwatini (Fig. 1). By combining demographic data and trends available (INE 2018) with direct observations on the ground, the map shows the unbalanced and fragmented urbanization processes involving the different districts and urban areas of the province, where the Greater Maputo represents the main epicentre of urbanization and metropolization dynamics.

A series of surveys of this territory has indeed highlighted a punctual and discontinuous distribution of settlements and small villages along the main roads among small semi-urban centres, the railway and the (major and minor) rivers of the province: a combination of dispersion and agglomeration factors of urban activities (Macucule 2016: 205) seems to be at play with different degrees and outcomes.

The second map (Fig. 2) focuses on the current and future/planned urbanization patterns, expected to expand mainly north–south, enclosed between the coastline with its wet inland strip (east) and the mountainous landscape in the west (corresponding to the national border), but also slowed down by the limits of the Umbeluzi (south) and Incomati (north) Rivers.

The map also makes evident what Macucule (2016: 203) defines as "proto-metropolisation (…)" synthesizing the primary socio-economic dynamics which are shaping it:

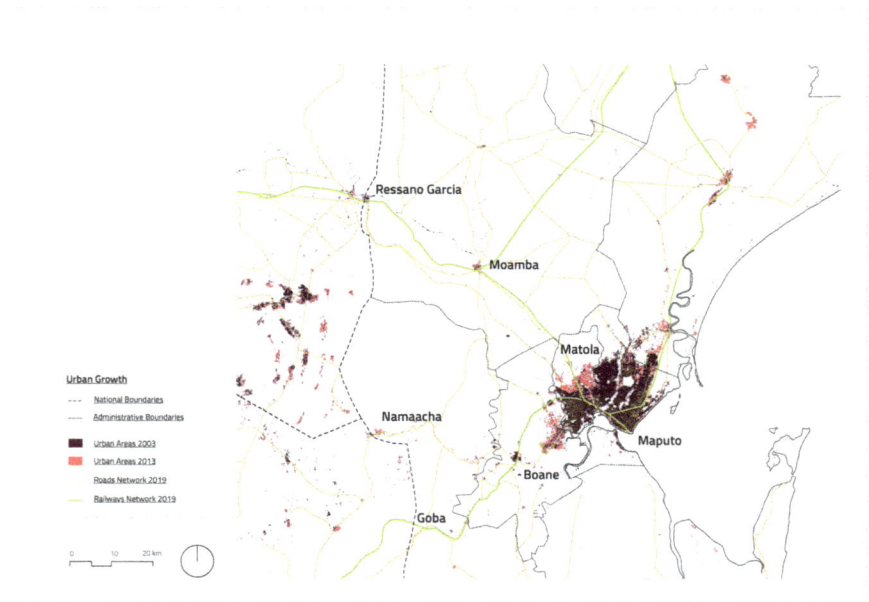

Fig. 2 Urban growth and infrastructure Elaboration by "Boa_Ma_Nhã, Maputo!" team 2020 (A. Buoli, A. Frigerio)

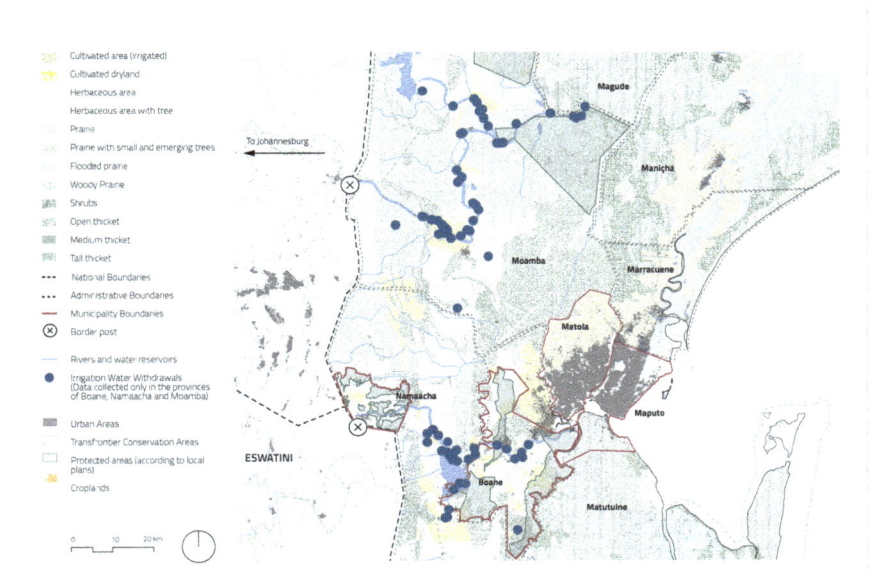

Fig. 3 Land uses, ecosystems and landscapes Elaboration by "Boa_Ma_Nhã, Maputo!" team 2020 (A. Buoli, A. Frigerio, A. Macchiavello)

the increase in the household income and in the use of cars, stimulating the option to increasingly live away from the centre (in Matola Rio, Boane and Marracuene), the large and multinational industries (Mozal, large retail and industrial markets), the increasingly strong connection of the local markets with South Africa (the improvement of the Maputo Development Corridor, the construction of the Maputo-Witbank motorway).[6]

Additional factors of metropolization are related to "international migration of people and capital linked to the advent of mineral resources in Mozambique and strong real estate investments (housing, commerce and services) in the centre which, together with the construction of the new ring road, are in- creasing the polarizing potential of the city of Maputo and the reach of the AMGM [Greater Maputo Metropolitan Area]" (Macucule 2016: 203).

3.2 Land-use Ecosystems and Landscapes/Hydrology and Croplands

The following two maps (Figs. 3 and 4) combine the current land uses, water withdrawals and hydrologic network; the perimeter of the protected areas of the three

---

[6] Original text in Portuguese, translated by the author.

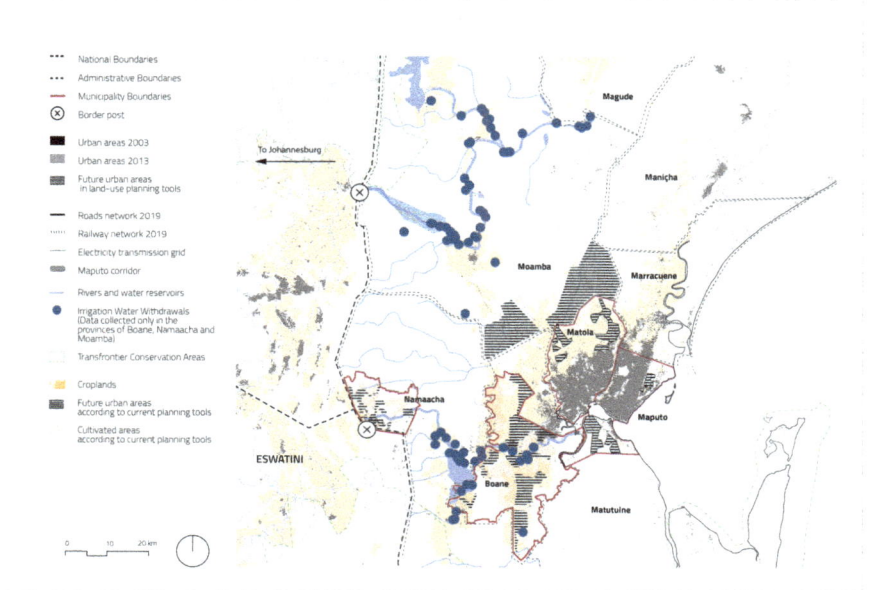

Fig. 4 Planned urban development and cropland erosion Elaboration by "Boa_Ma_Nhã, Maputo!" team (A. Buoli, A. Frigerio, A. Macchiavello)

districts of Moamba, Boane and Namaacha; and the municipalities of Maputo and Matola. The map was produced by combining cartographic information from the "District profiles" (República de Mozambique 2014a, b, c) produced in 1999 by CENACARTA and published by the Province of Maputo, with data extracted from the Global Forest Watch database, and data provided by the Ministry of Agriculture and Food Security of Mozambique on the irrigation schemes along the Umbeluzi and Incomati River (Boa_Ma_Nhã, Maputo! 2020a).

In these images, the Maputo metropolitan area appears as a complex mosaic of different landscapes, agroecological systems, infrastructure networks and urbanization patterns strongly determined by its geographical features. Topography, geology and hydrology set the framework for the potential development of the Maputo metropolitan urbanization and represent basic resources for the small- and medium-sized urban centres in the area.

From an environmental point of view, the image also shows how forests have almost disappeared in the region, despite planning tools promoting reforestation initiatives through community engagement, such as the Mozambique Forest Law of 1997 and the National Policy and Strategy for the Development of the Forestry and Wildlife Sector. Direct observations on-site also allowed to get a hint of the extent of mining and excavations occurring in the area, with little respect towards ecological assets.

The map also depicts the hydrography and agricultural system in the region, including Maputo's green belt, which expands north towards the Marracuene District and along the Incomati River. This image, together with the previous map, shows the dependency of agriculture production from water sources (two rivers Umbeluzi and Incomati and the related irrigation schemes) which are fundamental for the fertile irrigated lands. These will be more and more disputed to be used for cash crop and small-scale agriculture, industrial production and mainly by low-density urban "sprawl" settlements.

Planning tools define protected ecological corridors and designate farming areas, but the effectiveness is limited due to physical discontinuity and lack of awareness and implementation. Land and water competition across the province are and will be one of the crucial challenges in the future of Maputo and vast areas of Boane, Moamba and Namaacha Districts.

3.3 Planned Development and Cropland Erosion

The last map (Fig. 4) combines the data presented in the previous cartographies and synthesizes the main trends and planning provisions regarding urbanization patterns, from one side, and the loss of productive and fertile lands for agriculture and the continuity of agroecological systems.

Planning fragmentation clearly emerges as a crucial issue in this map, revealing territorial pockets—rapidly urbanizing in between planned areas—that are still not covered by clear and compelling planning indications (such as the Matola Rio sub-district, between Matola and Boane Municipalities) and the centripetal urbanization trends around the conurbation of Maputo-Matola. As a matter of example, the District Land Use Plans (PDUT) of Moamba foresees urban growth not in relation to the main urban centre in the district (Moamba City), but instead along and towards the border with the municipality of Matola that emerges as an attractor of urbanization patterns and migratory dynamics. Moreover, in most cases, urbanization occurs by fencing plots and limiting farming opportunities and failing to plan public and ecological spaces that could rebalance the urban footprint.

On the other hand, the PDUT of Boane suggests an expansion of the main urbanized areas north to south along the central infrastructural axis connecting the district to Maputo, but at the same time overlapping with the existing fertile lands providing more accessible and local food to a fast-growing population. This image makes explicit the existence of problematic governance issues due to the inconsistency of administrative borders with ongoing and future metropolitan dynamics and challenges, on the one hand, and the need to safeguard the persistence of key natural resources and productive agricultural landscapes.

4 Concluding Remarks and Open Prospects

This chapter presented some methodological and operative insights from "Boa_Ma_Nhã, Maputo!" research project to deal with the production of meaningful and *usable knowledge* frameworks in a context of low accessibility and availability of territorial data and information. By overlapping the complex scenario of existing planning documents and approaches for the Maputo Province with current conditions and trends, and, at the same time, designing multi-dimensional cartographic tools, we aimed at showing the possibility of operating a robust and effective spatial analysis in such conditions, based on the integration of different data sources, sometimes geo-referenced or based on official surveys, most often drawn on direct observations on the field and multi-dimensional interpretations. By embracing uncertainty and (a certain level of) approximation as a (productive) side effect of a research-by-design methodology, the project showed the need to move from a sectorial and "top-down" approach to spatial analysis to one in which different views and voices are included in the construction of a kaleidoscopic cartographic operative framework which can open to other visions and images of change for the Maputo Province.

The main result of this operation regards a more integrated metropolitan framing to envision a territorial strategy in which issues of more sustainable land-use provisions and the protection and long-term management of natural assets are embraced and supported to produce eco-systemic services to the whole region. This has been structured by focusing on regional geographic features in terms of continuous blue and green infrastructures identified in territorial connections to be better planned and defended. In this perspective, "Boa_Ma_Nhã, Maputo!" team has further investigated and spatialized the main WEF trends, opportunities and challenges to draft pilot projects proposals, strategic scenarios and trans-scalar policy recommendations. These are presented in the following contributions to the volume, providing practical applications of the proposed research-by-design methodology, a new and inventive way of looking at the Maputo metropolitan region to suggest other possible ways to envision its future.

Acknowledgements The research results and cartographic materials presented in this chapter have been produced in the context of the "Boa_Ma_Nhã, Maputo!" project, in collaboration with Alessandro Frigerio, Laura Montedoro and Alessia Macchiavello, who contributed designing the maps presented in this essay.

References

Amadife EN, Warhola JW (1993) Africa's political boundaries: colonial cartography, the OAU, and the advisability of ethno-national adjustment. Int J Polit Cult Soc 6:533–554. https://doi.org/10.1007/BF01418258

de Araújo MGM (2006) Espaço urbano demograficamente multifacetado: As cidades de Maputo e da Matola. Available via APDemografia: http://apdemografia.pt/files/1853187958.pdf. Accessed 15 November 2021

Boa_Ma_Nhã (2020a) Assessment Report. Politecnico di Milano, Milan

Boa_Ma_Nhã (2020b) Planning Tools Report. Politecnico di Milano, Milan

Instituto Nacional de Estatistica (INE) (2018) Anuario Estatistico 2017 - Moçambique. INE. Available via INE. http:/www.ine.gov.mz/estatisticas/publicacoes/anuario/nacionais/anuario-est atistico-2017.pdf. Accessed 15 November 2021

Foucher M (2020) African Borders: Putting paid to a myth. J Borderlands Stud 35(2):287–306. https://doi.org/10.1080/08865655.2019.1671213

Frigerio A (2020) Mapping: questioning boundaries, rights and sustainable development in the west Nile region, Uganda. In: Gaeta L, Buoli A (eds) Transdisciplinary views on boundaries towards a New Lexicon. Fondazione Giangiacomo Feltrinelli, Milan, pp 132–216

Jenkins P (2012) Home space: context report. Available via the Danish Council for Independent Research. http://homespace.dk/tl_files/uploads/publications/Full%20reports/HomeSpace_Context_Report.pdf. Accessed 15 November 2021

Heinrigs P (2020) Africapolis: understanding the dynamics of urbanization in Africa. Field Actions Science Reports 22:18–23

Macucule D (2016) Processo-forma urbana: Restruturação urbana e governança no Grande Maputo. Dissertation, New University of Lisbon. Department of Geography and Regional Planning

Mbembé A, Rendall S (2000) At the edge of the world: Boundaries, territoriality, and sovereignty in Africa. Publ Cult 12(1):259–284

Michalopoulos S, Papaioannou E (2011) The long-run effects of the scramble for Africa. National Bureau of Economic Research, NBER. Available via NBER. http://www.nber.org/papers/w17620. Accessed 15 November 2021

Montedoro L, Buoli A, Frigerio A (2020) Towards a metropolitan vision for the Maputo province. Maggioli Editore, Santarcangelo di Romagna

OECD and Sahel and West Africa Club (2020) Africapolis, mapping a new urban geography Sahel and West Africa club. Available via OECD. https://www.oecd.org/publications/africa-s-urbanisat ion-dynamics-2020-b6bccb81-en.htm. Accessed 15 November 2021

Piscitelli P (2018) Mobile urbanity. Translocal traders and city in Southern Africa. Planum Publisher, Rome-Milan

República de Moçambique (2007) Lei n. 19/2007 - Lei de Ordenamento do Território. Maputo

República de Mozambique. Ministério da Administração Estatal (2014a). Perfil do Distrito de Boane. Provincia de Maputo. Maputo

República de Mozambique. Ministério da Administração Estatal (2014b). Perfil do Distrito de Moamba. Provincia de Maputo. Maputo

República de Mozambique. Ministério da Administração Estatal (2014c). Perfil do Distrito de Namaacha. Provincia de Maputo. Maputo

Scotto G (2020) Infrastructures, borders, and the making of the African territory: the case of Zambia. In: Gaeta L, Buoli A (eds) Transdisciplinary views on boundaries towards a New Lexicon. Fondazione Giangiacomo Feltrinelli, Milan, pp 112–131

Watson V (2009) 'The planned city sweeps the poor away…': Urban planning and 21st century urbanization. Prog Plan 72:151–193

Energy-Food Challenges and Future Trends in Mozambique and in the Maputo Province

Lorenzo Rinaldi and Davide Danilo Chiarelli

Abstract Mozambique is one of the richest countries in sub-Saharan Africa in terms of natural resources and agro-ecological assets. However, the country does not look on track for the achievement of the SDG 2 (Zero Hunger) and 7 (affordable and clean energy) by 2030. The population with access to electricity is around 30%, while 1.9 million people are estimated to be in high levels of acute food insecurity, with a strong imbalance between urban and urban areas. The Maputo Province, being the economic centre of the country, faces an above-average condition in respect to other areas of Mozambique. The policy frameworks as regards both energy and food production are indeed mainly focused on the development of the Great Maputo area. At the same time, major weaknesses hindering a more sustainable and even access to basic resources in this area are related to a) the ongoing use of coal and biomass for cooking purposes and the difficulties in adopting more clean and renewable energy sources; b) the impact of climate change, water scarcity and land / water competition on small-scale subsistence farming, as well as land-grabbing issues related to the presence of transnational food companies. In this context, the main challenge for policymakers is to translate the richness and availability in resources into effective policies, to support local economic activities and enable sustainable development also in terms of clean energy, and healthy and nutritious food access. Based on insights from "Boa_Ma_Nhã, Maputo!" project, this chapter focuses on these two interrelated dimensions of the WEF nexus (Energy and Food), focusing on specificities of the Great Maputo area. The essay further discusses future patterns provided by authoritative institutions to provide insights on potential criticalities and challenges to be overcome and orient more integrated and sustainable energy and food security-related policies.

L. Rinaldi (✉)
Department of Energy (DENG), Politecnico Di Milano, Milan, Italy
e-mail: lorenzo.rinaldi@polimi.it

D. D. Chiarelli
Department of Civil Engineering (DICA), Politecnico Di Milano, Milan, Italy
e-mail: davidedanilo.chiarelli@polimi.it

1   Overview of the Energy Situation in Mozambique and the Maputo Province

1.1   Electricity Production and Access

Mozambique is a country full of natural resources, representing a valuable opportunity if effectively exploited to produce energy. The Mozambican electricity production mix is poorly diversified: the main source for electricity production is water, which is exploited for 83% of the domestic production (IEA 2019), thus creating competition with food production. In 2015, it was estimated that 73% of water withdrawals were for agricultural purposes and 65% for irrigation. However, crop production remains mainly rainfed in the country (FAO 2016). Mozambique is crossed by 104 river basins including the Zambesi River, the fourth longest river in Africa, and the Limpopo, the second main river of the Southern Africa region. The main hydroelectric plant in operation is represented by the 2075 megawatt (MW) of Cahora Bassa, located in the Tete Province, at the border with Zambia, Zimbabwe and Malawi. Due to this strategic geographical location, Cahora Bassa represents a cardinal electricity production node, and therefore a large portion of its production is exported among other Southern Africa Power Pool (SAPP) member countries. In matter of electricity supply from hydroelectric plants, very few is the installed capacity in the Great Maputo area, which hosts only the plant of the Corumana dam (16 MW) located in the Moamba District. In terms of future development, more than 1400 hydroelectric projects are currently in the feasibility assessment or under development. One of the projects with a major priority is the Mphanda Nkuwa plant with a planned capacity of 1500 MW. As a whole, the Mozambican hydropower potential is estimated to be 19 GW (Gesto-Energia 2000).

In the second position within the electricity production mix, there is natural gas, accounting for almost the totality of the remaining domestic electricity, despite the huge potential for exploitation of other sources. Natural gas represents one of the Mozambican main assets from a strategic point of view, placing the country in a relevant position on the international market, especially towards Asia. Two are the main natural gas reserves off the coast of the country. The Pande and Temane gas reserves, located in southern Mozambique, are estimated to contain 3.5 trillion cubic feet of natural gas. Developed by the South African Sasol, 90% of their production is exported to South Africa via pipelines while the remaining is left for domestic use, exploited in the cities of Maputo and Matola (MIREME 2018).

The second reserve is the Rovuma Basin gas fields, off the coasts of the Cabo Delgado Province, in the North of the country. The recent discoveries of the Rovuma Basin represent some of the largest gas fields in the world with a capacity of 128 trillion cubic feet (EDM 2018). This massive potential is envisioned on one side to increase the power generation capacity through the implementation of combined cycle power plants, and on the other to boost export revenues from the sector.

The national power system is managed and operated by Electricidade de Moçambique (EDM). According to the company statistics (EDM 2015), the national electricity grid nearly doubled in length from 2009 to 2015, increasing from 9252 km to 16,662 km. The southern provinces, in particular, the Great Maputo area, benefit of a much better electric service with respect to the rest of the country, and this is both a cause and a consequence of the higher level of development. The grid extension trend was uniform within the country, yet Mozambique is still at the bottom line in terms of electricity infrastructure. Such weak infrastructures lead to one of the lowest rates of access to electricity in the world. More than 70% of the population lacks access to electricity; moreover, this value increases dramatically considering rural areas only, reaching almost 95% (World Bank 2021). Access to electricity is also strongly dependent on the region in Mozambique, with an alarming imbalance from North and Centre Provinces (around 17%) to South (56%). The Maputo area is the most developed region in the country in terms of electricity access: 91% of the population of the capital was connected to the national grid in 2015, while, in the surrounding districts, this value approaches to 79%.

1.2 Biomass: The Backbone of the Energy Mix

According to the 2007 census data, only 29% of the households in the Maputo Province reported electricity as their main source of energy supply. However, more than 35% of the population of the area was connected to the national grid. With relation to the total population of that year, it means that 15 thousand households in the area preferred other energy sources to electricity. Considering this data refers to the most developed region of the country, it is reasonable to derive a far worse condition in the other regions.

Figure 1 highlights a strong dependence on biomass: aside from representing the second resource in terms of domestic energy production (left-side chart), it is also not exported at all. On the right-side chart, it is clear how biomass is the backbone of Mozambican energy consumption, especially from households and public and commercial services. Everywhere in Africa, biomass is the heart of the energy mix, and due to economic but also cultural reasons, it will be difficult to replace. Despite significant growth in GDP per capita from 611 to 1258 USD between 2000 and 2018, according to the IEA, in 2012, about 25 million people in Mozambique were relying on the use of traditional solid biomass for cooking purposes, especially in rural areas (IEA 2014). Moreover, in the same period, only an additional 2% of the population gained access to clean cooking solutions (IEA 2019).

A focus on the districts of Boane, Moamba and Namaacha is also provided in Table 1. According to data coming from the INE (National Statistics Office), despite oil products were reported to be the primary energy supply source at domestic level, with more than 60% of share in all Boane, Moamba and Namaacha Districts, electricity is already the second most utilized source with around 20% share (INE 2013a, b, c). Despite such conditions can be attributed, in absolute terms, to a underdeveloped

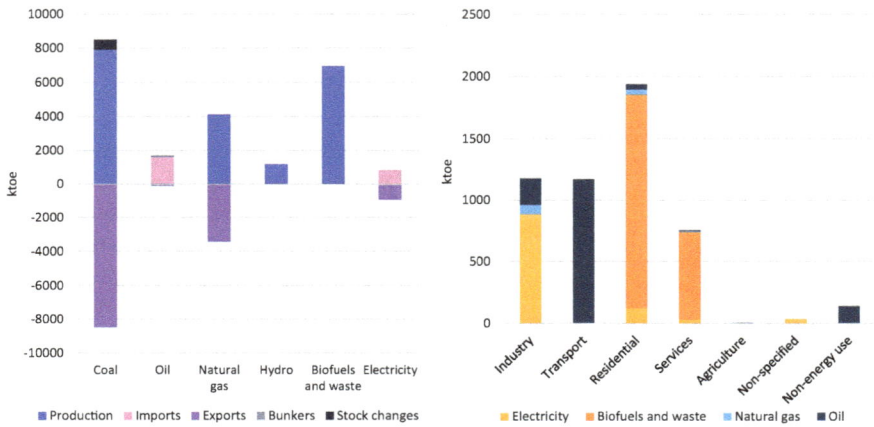

Fig. 1 Energy supply (left) and final consumption (right) in Mozambique, 2018. *Source* Elaboration of the authors based on IEA data (IEA 2021)

region, it is to be noted that in the rest of the country the numbers may be much more shifted towards less clean fuel utilization.

Still referring to the Great Maputo area, aside of the residential demand, the only industrial utility to be served is the Mozal smelter, which accounts for a large portion not only of the local, but also of the national demand of electricity. Electricity to serve the Mozal is coming from exports, since there is no infrastructure that links it with the main power plants located in the northern provinces.

1.3 Unexploited Opportunities for Improvement

Despite the unfortunate current energy situation, there are considerable margins for improvement. Starting with the hydroelectric potential, the country is crossed by 104 river basins of which 13 are main basins counting the Zambesi River, the fourth longest river in Africa, and the Limpopo, the second main river of the South African region. More than 1400 projects are identified to increase the generation capacity: the one with the highest priority is the Mphanda Nkuwa plant with a planned installed capacity of 1500 megawatt (MW). In the Maputo area, another hydroelectric plant of 15 MW capacity is under construction in the Moamba District. According to the Mozambique Renewable Energy Atlas, the Mozambican hydropower potential is estimated to be 19,000 MW (Gesto-Energia 2000). A feasibility study is underway for the construction of a biomass power plant in Moamba, with a capacity of 30 MW. Other relevant sources of unexploited potential are solar and wind energies, which are estimated as 23 terawatt (TW) and 4.5 gigawatt (GW), respectively. A 30 MW wind power plant is expected to be in operation by 2027 in Namaacha. Considering the currently installed capacity of 1 GW for photovoltaic, this provides insights on the big

Table 1 Households preferred energy sources in Moamba, Boane and Namaacha. *Source* Elaboration by the authors based on data from INE (2013a, b, c)

	Boane		Moamba		Namaacha		Maputo Province		Maputo City	
	n. HH	%	n. HH	%	n. HH	%	n. HH	%	n. HH	%
Electricity	6086	24.0%	2674	18.3%	2302	21.7%	79,054	29.3%	137,756	63.0%
Generator/solar panel	94	0.4%	49	0.3%	51	0.5%	1012	0.4%	391	0.2%
Gas	13	0.1%	5	0.0%	2	0.0%	131	0.0%	81	0.0%
Oil products	16,245	64.0%	9777	66.9%	6408	60.5%	161,738	59.9%	64,192	29.4%
Candles	2564	10.1%	1405	9.6%	1382	13.1%	21,998	8.1%	15,164	6.9%
Batteries	41	0.2%	21	0.1%	15	0.1%	703	0.3%	533	0.2%
Wood	265	1.0%	600	4.1%	353	3.3%	4368	1.6%	135	0.1%
Others	85	0.3%	79	0.5%	75	0.7%	920	0.3%	272	0.1%
Total households	**25,393**		**14,610**		**10,588**		**269,924**		**218,524**	

steps that may be performed in this direction. Three 30 MW photovoltaic plants are expected to be deployed in Boane, respectively, in 2024, 2030 and 2038. However, mainly due to the elevated cost of renewable technologies, their deployment is not highlighted as a priority in the government masterplans, which are mainly focused on the development of the gas extraction and export sectors. The foreseen investments in the energy sector are concentrated, in fact, towards the development of the fossil fuels (specifically gas) industry, which may represent a rapid way to place Mozambique in a better position on the energy market.

Another aspect that deserves a specific discussion is the off-grid potential. As mentioned previously, the national grid is sparse, and its improvement is highly capital-intensive. Exploiting the off-grid potential is of paramount relevance for poli-cymakers. However, providing access to electricity is not just a matter of connecting the user to a smart meter, but ensuring a durable, reliable and affordable service (Bhatia and Angelou 2015). Usually, the most adopted energy solutions to provide access to electricity without relying on the national grid are solar home systems and microgrids. While the former is usually capable of serving a few basic appliances, the latter can provide proper access to electricity in remote regions, especially in the case of hybrid plants, in which renewable potential is complemented with storage technologies as well as diesel generators as back-up. While microgrids are widely diffused in Southeast Asia, they are not so common in Africa, and Mozambique is not an exception (ESMAP 2019). While the number of publicly financed projects is increasing, mainly managed by the Energy Fund of Mozambique (FUNAE), a sub-division of the MIREME dedicated to rural electrification projects, privately funded projects usually crash against a wall of difficult administrative barriers, which disincentivizes private stakeholders.

2 National Energy Policies in Mozambique

The government strategies in the matter of energy sector development are clear. Both the "Natural Gas Master Plan" and the "Integrated Master Plan of Mozambique Power System Development" state that the main ambition of the country is to increase the revenue coming from natural gas resources exploitation (MIREME 2018; INP 2014), having the long-term objective of improving social and economic conditions of the population.

Focusing on the power system, the government strategies comprise a develop-ment plan of the generation, transmission and distribution infrastructure for the time horizon 2018–2043. At the current state, the Mozambican national grid is divided into two independent systems: the southern and the central and northern systems. For this last reason, the generation development plan was developed in two stages: the first stage considers the two systems as separated and is implemented for the period 2018–2028, while the second stage considers the two systems as interconnected between 2029 and 2043, forecasting such link to occur in 2029.

The generation development plan targets all the on-grid power systems across the and is based on a least-cost method modelling approach. Indeed, since the country, as mentioned, has a huge potential in primary energy, the strategy proposed by the master plan is going to be implemented by using different kinds of power plants such as hydroelectric power plants, coal- and gas-fired power plants, solar and wind power units: adopting different technologies, different levels of investments may be needed. Given the current economic situation of the country, adopting a least-cost approach seems the most reasonable methodology to track the future deployment of capacity and fulfil the estimated demand for electricity, trying not to impact food production.

To estimate the future electricity demand, the final users were divided into three different categories: (a) *general customers* constituted by households and other small customers supplied by low voltage; (b) *medium-large customers* supplied by low and medium voltage and (c) *high-voltage customers* served by high volt. Another category is that of the *special customers*, which is composed of all the customers whose contract considers a supply of 1 MW or more. The demand forecast was performed at three levels: at the customer side, at transmission substation and a power station, considering transmission and distribution losses and was estimated based on the macroscopic indicators (i.e. GDP, population growth, etc.), complemented by specific analyses dedicated to the special customers since they impact significantly the overall demand. The correlation among electricity demand, population growth, GDP, electrification rate and electricity tariff has been analysed using historical data. The results showed that there was a strong correlation between the first three indicators and the demand, defining them suitable. On the other hand, regarding the electricity tariff, instead, since the correlation was weak, it was not significant for the forecasting.

The optimization algorithm takes as input the estimated demand and provides the best technological solutions to meet such demand level while minimizing the total investment and operation costs to be put in place. The installed units are selected among a set of already planned projects, and technical parameters are characterized for each generator unit.

2.1 Possible Criticalities in Future Trends

The policy documents foreseen a large deployment of capacity, estimated in almost 10 GW in total up to the year 2043, to be compared with the current installed capacity for power generation which is lower than 3 GW. The main deployed technologies are expected to be hydroelectric power plants, with almost 4.4 GW of new capacity, gas-fired power plants (2.6 GW) and coal power plants (1.9 GW). Renewable energy technologies shall keep covering a marginal portion of the total electricity demand, with an expected capacity of 680 MW in total. The strategy proposed by the Mozambican government does not differ significantly from the ones reported by other institutions. The IEA, for instance, in its "Africa Energy Outlook report", displays an electricity production mix up to 2040 which is strongly unbalanced towards hydro and gas plants, with a negligible share of renewable production (IEA 2019).

The planned technological solutions provided in the policy documents present, however, some critical issues attributable to the underlying modelling approach adopted. Firstly, the selected installed units come from a pre-defined list of already planned projects. This is in line with current policy statements, which aims at maximizing the exploitation of recently discovered natural gas reserves: gas is expected to increase its share in the electricity mix as well as its weight in foreign trades since in 2040 more than 50 of the 80 billion cubic feet (bcf) produced will be exported. However, the followed approach limits the degree of freedom of the model, which could provide alternative solutions, which may be more convenient.

Secondly, macroscopic trends such as population growth or cross-sectoral linkages are not accounted for within the boundaries of the model, and therefore the costs related to the resulting solution may neglect additional costs to be sustained in other economic sectors. The economic efforts, in any case, will be probably dramatically high: at least 110 billion USD may be necessary for the installation of the new plants and to improve the grid infrastructure. This translates into 5.5 billion USD to be invested each year, around 32% of the national GDP in 2014, the year of publication of the master plans. Considering the situation from a post-pandemic perspective, the challenge seems even more difficult, since the national economy experienced a downturn during the COVID-19 emergency.

In the end, it seems MIREME master plans do not focus on the growing trend of increasing access to clean cooking. In the most conservative scenarios, the share of the population gaining access to clean cooking solutions is expected to quadruple. Thanks to the widely available gas resources, the main resource in this matter is represented by the diffusion of Liquid Propane Gas (LPG) cooking stoves. However, nowadays, while 90% of energy needs for cooking are supplied by biomass and coal, the most adopted clean cooking technology is electric stoves over LPG ones (7% against 3% in the total energy for cooking supply mix). If this share will be respected in the future, it may represent a further issue that apparently has been neglected within the demand forecast process: electric cooking may significantly increase the consumption level of electricity, revealing the technological solution provided by the master plans as sub-optimal or even unsuitable.

3 The Agricultural System in Mozambique and in the Maputo Province

Mozambique is considered as having a great potential for agricultural production as well as energy production; however, it is not yet exploited nowadays. Malnourishment and food security are still on the agenda in the country.

Agriculture contributes 26% to the GDP and is the backbone of the country's development with about 70% of the population heavily dependent on agriculture as their primary source of livelihood (FAO 2016; Ferrão et al. 2018). The smallholder "family" sub-sector accounts for about 98% of the area under production: production is largely rainfed and with low yields. Food crops include mainly maize, cassava, rice and beans for household consumption. Private companies represent the remaining 2% of the cropland area, often irrigated, producing mainly cash crops such as cotton, cashew, sugarcane, tobacco and soybean (FAO 2016; Cammaer 2016; Silici et al. 2015).

Mozambique has considerable agricultural potential thanks to its rich endowment of natural resources. In this context, only 90,000 ha are irrigated, even though irrigation potential was estimated to be around 3.1 million ha (de Sousa et al. 2017).

In the Great Maputo area, for example, approximatively 70% of the maize water requirement, the most widespread crop in the country, is satisfied by precipitation during the wet season. In the district of Moamba water gap is higher only 67% of crop water demand is satisfied by rainfall, while in Namaacha we reached 72.3%. Thus, we can expect that the total crop yield could not be reached unless water with irrigation is provided in the field. Water withdrawal for agricultural purposes in the artificial reservoir of the Pequenos Libombos dam, along the Umbeluzi River, clearly highlighting how the water withdrawal for agriculture mainly occur during the dry season.

Annual water scarcity map (Fig. 2) shows how water is generally available in the district of Namaacha, while water scarcity is experienced in the district of Boane and Moamba, thus showing how proper irrigation schemes could be realized without compromising local availability of water.

3.1 Possible Future Trends and Guidelines

Crop diversification and organic production have been highlighted by local expertise[1] as a way to enhance small-scale agriculture in the region. These suggestions are identified in order to provide more resilient strategies for production that is less connected with the local market of a single product and could be at the same time competitive with the neighbouring production coming from eSwatini and South Africa. Thanks to irrigation and cheaper production cost, products from South Africa are sold at a lower price than the same products harvested in Mozambique, where moreover small farmers are only able to get one unique harvest during the wet season.

Improving irrigation system is a necessary strategy to increase crop production of local farmers ensuring a double cropping during the year. Irrigation systems are usually possible when farmers organize themselves in associations or cooperatives in order to ensure their maintenance, while the initial financial input is anyhow

[1] Empirical evidence has been provided—in the context of "Boa_Ma_Nhã, Maputo!" project—thanks to the discussion with local farmers and researchers, as well as direct observation of farming activities in Boane and Namaacha. Interviews included members of the 25 de Septembro cooperative in Boane and colleagues from the Sabie Agricultural Research Centre, Department of Agronomy and Forestry Engineering of the Mondlane University.

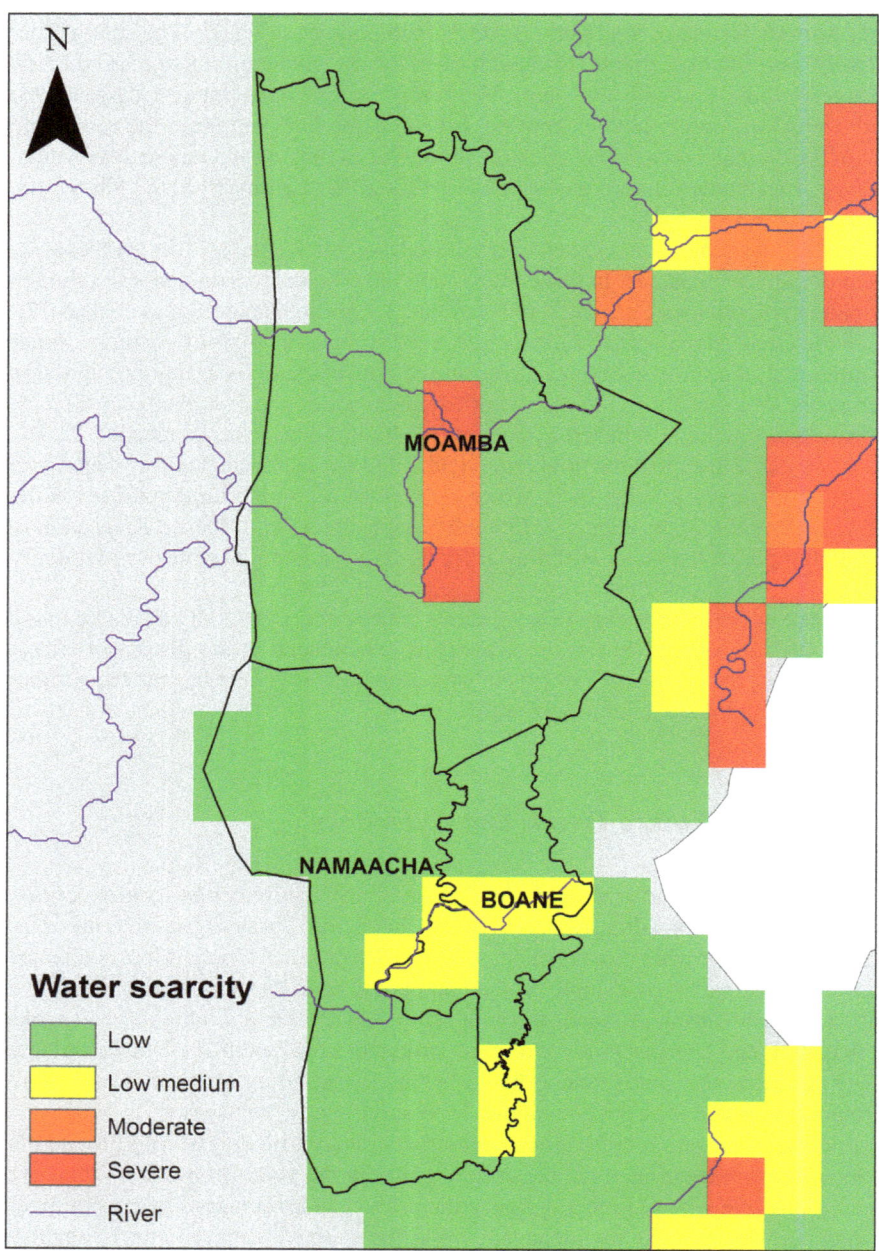

Fig. 2 Water scarcity map in the Maputo province. *Source* Elaboration by "Boa_Ma_Nhã, Maputo!" team (D. D. Chiarelli), 2019

provided by external private companies or international cooperation agencies. An example is represented by the 25 de Setembro cooperative in Boane that is currently including more about 35 farmers mainly harvesting horticultural, vegetables and maize. Improving the knowledge of local farmers is another important key point to be stressed. Agrarian schools and universities are present in the Maputo Province, but usually diplomats and graduate students prefer administrative careers in local institutions over entrepreneurial careers as farmers, thus without a return of acquired knowledge on the ground.

3.2   Foreign Investment in the Agricultural Sector and Consequences

To have a complete overview of agricultural production is Mozambique, however, we cannot neglect to include foreign investment in the agricultural sector. Currently, in Mozambique there is no private ownership of the land, everything is state-owned. With the 1997 Land Law, the State declared itself owner of the land but committed itself to guarantee the right to use it. This right is perpetual for those who make it a domestic and subsistence use, while companies in search of land to be dedicated to agribusiness are required to renew this right, at low cost, every 50 years, provided that the use complies with approved development plans and environmental constraints.

Because of the nationalization of land, a weak and unstable political situation, and especially because of the liberal and market-oriented development model promoted by the central government, Mozambique is heavily subjected to land acquisition mechanisms for private investments in agribusiness, forestry and mining. These mechanisms are creating serious impacts both on the environment and smallholder farmers (Rulli et al. 2018; Chiarelli et al. 2021).

Moreover, the rare irrigation schemes are often associated with cash crops like bananas and sugarcane mainly harvested by foreign investors, while smallholder farms usually survive thanks to rainfed agriculture. Examples of these foreign investments are Bananalandia, located in the Boane District, and Agrisol, in the Moamba District. In these districts, water gap is high, only 67–73% of crop water demand is satisfied by rainfall. Thus, only providing water with irrigation can help reach total crop yield. These harvested areas also receive high inputs in terms of fertilizers and pesticides.

4   Conclusions

Mozambique is undergoing an unfortunate situation, characterized by general underdevelopment, poverty, malnutrition and difficulties in access to food which is reflected also in the energy sector. Even if the Maputo metropolitan area is by far the most developed region of the country, the current infrastructure is not sufficient to satisfy basic needs and planned energy and food security solutions for the next two decades present important criticalities which should be investigated and tackled carefully. Despite the current conditions are not promising in the perspective of achieving SDG 2 and 7 targets by 2030, there are significant margins for improvement, from the unexploited renewable potential to the diffusion of off-grid microgrids for energy supply, while the revenues coming from the natural gas market may enable positive dynamics of economic growth. The agricultural sector needs also new strategies to increase crop production for both local and international markets. Improving irrigation schemes could be the best way associated with an increase in energy supply if local farmers may also have access to them.

References

Bhatia M, Angelou N (2015) Beyond connections energy access redefined. ESMAP Technical Report. Available via World Bank. https://openknowledge.worldbank.org/handle/10986/24368. Accessed 22 September 2021

Cammaer R (2016) Tracing sustainable agriculture in Mozambique: from policy to practice. International Institute for Environment and Development Working Paper Series. IIED, London

Chiarelli DD, D'Odorico P, Davis KF, Rosso R, Rulli MC (2021) Large-scale land acquisition as a potential driver of slope instability. In: L. Degrad. Dev. 32(4):1773–1785 https://doi.org/10.1002/ldr.3826

de Sousa W et al (2017) Irrigation and crop diversification in the 25 de Setembro irrigation scheme. Mozambique. Int. J. Water Resour. Dev. 33(5):705–724. https://doi.org/10.1080/07900627.2016.1262246

EDM—Electricidade de Mocambique (2018) EDM strategy 2018–2028. EDM, Maputo

EDM—Electricidade do Moçambique (2015) Relatorio anual estadistica. EDM, Maputo

ESMAP—Energy Sector Management Assistance Program (2019) Mini grids for half a billion people: market outlook and handbook for decision makers. ESMAP Technical Repor. Available via World Bank: https://openknowledge.worldbank.org/handle/10986/31926. Accessed 22 September 2021

FAO—Food and Agriculture Organization (2016) AQUASTAT country profile—Mozambique. FAO, Rome

Ferrão J, Bell V, Cardoso LA, Fernandes T (2018) Agriculture and food security in Mozambique. J Food Nutr Agric 1(1):7–11. https://doi.org/10.21839/jfna.v1i1.121

Gesto-Energia SA (2000) Renewable energy atlas of Mozambique. Gesto-Energia, Alges

IEA—International Energy Agency (2021) IEA data and statistics. Available via IEA. https://www.iea.org/data-and-statistics. Accessed 22 September 2021

IEA—International Energy Agency (2019) Africa Energy Outlook. Available via IEA. https://iea.blob.core.windows.net/assets/2f7b6170-d616-4dd7-a7ca-a65a3a332fc1/Africa_Energy_Outlook_2019.pdf. Accessed 22 September 2021

IEA—International Energy Agency (2014) Africa Energy Outlook. A focus on the energy prospects in sub-Saharan Africa. Available via IEA. https://www.icafrica.org/fileadmin/documents/Knowledge/Energy/AfricaEnergyOutlook-IEA.pdf Accessed 22 September 2021

INE—Instituto Nacional de Estadísticas (2013a) Estadísticas do Distrito de Boane. INE, Maputo

INE—Instituto Nacional de Estadísticas (2013b) Estadísticas do Distrito de Moamba. INE, Maputo

INE—Instituto Nacional de Estadísticas (2013c) Estadísticas do Distrito de Namaacha. INE, Maputo

MIREME—Ministério dos Recursos Minerais e Energia (2018) Integrated Master Plan—Mozambique Power System Development. MIREME, Maputo

Republica de Mozambique. Instituto Nacional de Petróleo (2014) Natural Gas Masterplan. Available via INP: http://www.inp.gov.mz/en/Policies-Legal-Framework/Policies/NATURAL-GAS-MASTER-PLAN2. Accessed 22 September 2021

Rulli MC, Passera C, Chiarelli DD, D'Odorico P (2018) Socio-environmental effects of large-scale land acquisition in Mozambique. In: Bellaviti P, Petrillo A (eds) Sustainable urban development and globalization. New strategies for new challenges—with a focus on the global south. Springer, Cham, p 377–389

Silici L, Bias C, Cavane E (2015) Sustainable agriculture for small-scale farmers in Mozambique. Country report. International Institute for Environment and Development, London

WB—World Bank (2021) Wolrd Bank open data. Available via Word Bank. https://data.worldbank.org/. Accessed 22 September 2021

Trans-scalar and WEF-sensitive Strategic Scenarios for an Integrated Territorial Development. A Proposal for the Maputo-Boane-Namaacha Transect as a Green-Blue Metropolitan Armature

Alessandro Frigerio

Abstract The chapter presents some insights from "Boa_Ma_Nhã, Maputo!" research project concerning the need to reconceptualize spatial planning in the specific context of a growing sub-Saharan African city coping with the effects of climate change and a fragmented planning framework. The first part discusses the opportunity to embrace a metropolitan perspective and investigate alternative strategies and tools to overcome the weakness of mainstream planning paradigms when dealing with WEF nexus challenges in such a context. The second part presents an experimental study and proposal for the Maputo Province, including four pilot projects, one of which is discussed as an example of how trans-scalar strategic scenarios could facilitate WEF-sensitive metropolitan governance and planning processes: the Maputo-Boane-Namaacha (MaBoNa) Transect. This is envisioned as an integrated territorial green and blue infrastructure with the potential to set a synergy among public, private and third sector actors in supporting the sustainable development of the region, thanks to better balanced rural-urban linkages. From territorial guidelines to local development plans and projects, drafting strategic scenarios could mean co-producing platforms for building knowledge, promoting negotiation and facilitating decision-making: common grounds to foster cooperation.

1 Towards Alternative Planning Frameworks: The Need for a Metropolitan Perspective

Maputo metropolitan area extends far out its municipal boundaries. The rural-urban dynamics determining its urban development, investigated through a WEF nexus lens, extend to the whole Maputo Province and even beyond national borders (as discussed in detail in Chap. 5). One of the main goals of "Boa_Ma_Nhã, Maputo!"

A. Frigerio (✉)
Department of Architecture and Urban Studies, Politecnico di Milano, Milan, Italy
e-mail: alessandro.frigerio@polimi.it

research project[1] has been setting the framework of the spatial correlations between urbanization and WEF processes by crossing any super-imposed border, both in phenomenological and epistemological terms. Thus, pushing both spatial and disciplinary boundaries to re-draw a metropolitan image of Maputo by moving from its geographical structure and infrastructural system.

Overcoming the urban perspective, territorially limited to the city/non-city dichotomy, the metropolitan gaze extends to the territorial interlace of the metabolic flows that sustain a city's development, with a scale depending on the frame of reference used to analyse it (Dematteis 1997; Secchi 2000; Swilling 2013). Metropolitan systems always have a geographical rooting based on the availability of natural resources and are generally found on the ecotone line between two different ecological landscapes, where horizontal geographical and topographical networks (extension) intersect with vertical economic, social and ecological relations (depths) (Chiesa 2010), interweaving multiple inter-scalar patterns shaping built and unbuilt landscapes. The effort of reading, interpreting and taming these patterns and processes requires appropriate specific tools beyond traditional comprehensive and regulatory planning ones. As pointed out by Prosperi (2009), it seems that "a general theoretical model is needed: one that merges the disciplinary biased approaches of landscape ecology, intra-regional economic structure as expressed in the built environment, and governance structures".

In a spatial perspective, such metropolitan modelling efforts recall the synthetic concept of *Net-city* by Shane (2005), with metropolitan systems seen as a recombination of patches of natural and built environment, kept together by infrastructural armatures, the robust structural elements organizing processes of urban or rural transformation with different timing and various self-regulating patterns.

An operational effort in this sense, with specific regard to Global South contexts, has been carried out by Behrens and Watson, that in "Making Urban Places" (1996) elaborate on the concept of *Layout Planning Process* in the South African context, contrasting principles informing existing land-use-based planning tools. Urban development is seen as a process, and the purpose of layout planning is therefore presented as a way "to provide a spatial framework within which numerous collective and individual investments can be accommodated over time, in a mutually reinforcing and developmental manner", an "initiating and facilitative" tool to make infrastructural elements working together in a systemic way.

More recently, again with a Global South focus and with broader metropolitan perspective and adaptive character, Gouverneur (2014) works on this challenge through the concept of *Informal Armature*, intended as a design and managerial tool that strongly links performativity and morphological definition, with the aim of driving financial and human capital in an efficient way (in an age of global scarcity) and with the crucial quality of allowing transformation over time. To do this, Gouverneur redefines the basic elements of metropolitan systems (green and

---

[1] "Boa_Ma_Nhã, Maputo!" was a research project based at and funded by Politecnico di Milano and running between 2019 and 2020. A complete description of the project is provided in Chapter 4 of this volume.

grey armatures, nodes and patches), giving them specific tasks to perform in relation to their adaptive/inventive mission, but also describing the need for implementation patterns involving local communities and bottom-up stewardships.

The research work of these scholars clarifies how dealing with metropolitan systems in critical contexts, such as Maputo's one, implies re-defining analytical and planning tools to be trans-scalar, with a gradient of definition moving from strategic orientation (scenario, agenda, policies, guidelines) to operational implementation (projects) according to the scalar frame of reference. So, to be better said, a telescopic approach. Bernardo Secchi (2000) uses the metaphor of the telescope in the field of territorial studies, stressing how, by zooming in and out and changing the visual field, it is possible to get at the same time general tendencies and local details, thus understanding what links the ones with the others. Moreover, improving the understanding of these links requires the contribution of know-hows from different disciplines and a specific effort in highlighting the spatial correlations of the processes shaping the specificity of metropolitan territories, such as the ones related to food, water and energy cycles.

"Boa_Ma_Nhã, Maputo!", exploring the potential for alternative strategic planning tools for a sustainable growth of the Maputo Province, embraces that same telescopic and trans-disciplinary metropolitan perspective, investigating the role of trans-scalar and WEF-sensitive strategic scenarios to facilitate an adaptive and integrated territorial development by building on existing value assets and potential investments to be accommodated over time in appropriate synergy.

2 Territorial Armatures and the Role of Green and Blue Infrastructure

The strategic metropolitan paradigm introduced above sets a relevant change in the way of approaching the role of infrastructure. The dominant Anglo-Saxon rational-comprehensive planning model—largely diffused across the Global South due to colonial and post-colonial legacies and dynamics that are well explained by Abbott (2012)—still relies on the conventional concepts of *land-use* and *specialized infrastructure*, both unsuitable in contexts of rapid urbanization, scarce resources, and fragile governance. On the contrary, switching towards a metropolitan and strategic approach means introducing the ideas of (1) a gradient of planning determinacy (there are structural elements that should be planned as robust invariants and others that can be less regulated, to the extreme of leaving them self-regulating) and (2) multifunctional infrastructure (especially introducing the concept of green and blue infrastructure, able to provide multiple relevant socio-ecological services in an adaptive WEF perspective).

The metropolitan components are distinguished in continuous linear (green/blue-environmental and grey-transportation) and discontinuous (housing, productive, social patterns) systems. The way in which these elements are combined—and nodes articulated—in relation to the local geographical, historical and cultural context deter-mine the performance of the metropolis and its sustainability in a socio-ecological and economic perspective. Thus, strategic metropolitan planning should particularly take into consideration the territorial armatures providing for the robust structure orienting urbanization and the metabolic flows supporting it.

In this perspective, green and blue infrastructure is increasingly recognized as an important complement to conventional grey infrastructure in urban areas (European Commission 2015; Ramaswami et al. 2016; Pasquini and Enqvist 2019). The provision of services, in fact, is a key feature for territorial ecological armatures to perform as infrastructure. Green infrastructure or green and blue infrastructure (in a more comprehensive definition) can produce economic and environmental benefits and a range of ecosystem services provided with market value. Moreover, as stressed by Harrison et al. (2014), "in contrast to many grey assets, which are typically geared towards a single purpose, natural systems perform a range of functions". Ecological systems are "naturally multi-functional, simultaneously providing a suite of services such as food alleviation, cooling heat islands, carbon capture, water filtration, local food production and the provision of spaces for people and nature to reconnect (Roe and Mell 2012)" (Harrison et al. 2014). This encourages the protection, strengthening and maintaining of existing landscape networks and the implementation of green-grey blended solutions as backbone for sustainable urban development, an approach that expands the concept of *right to the city* (Lefebvre 1968), or combines it with the broader idea of *right to landscape* (Egoz et al. 2011). In Ortiz's words (2014), "balanced urban development begins with the recognition of a region's environmental heritage and a commitment to preserve and enhance it. [...] Preserving the ecological integrity of the metropolis requires a continuity of open, or unbuilt, space" that reconnects inner metropolitan parks with regional and national conservation areas. This role is particularly crucial in considering the urgency to set up solutions to mitigate or adapt to the effects of climate change.

The paradigmatic shift suggested by this theoretical perspective has a great relevance in the context of sub-Sharan African urbanism, "where the primary challenge is to integrate social development, and social equity, with environmentally sustainable practices" (Abbott 2012: 255). And this is especially because "the systemic resource-based approach to urban development radically changes the way in which land is both perceived and managed", with the priority "given to the social relationship between people and space, and not to the legal relationship between people and land" (Abbott 2012: 331). Thus, intending infrastructure as an economic, social and environmental driver to mediate urban resource flows could support a reconceptualization of planning as a less rational-comprehensive and a more strategic and flexible orienting and management tool, particularly promising to cope with contemporary challenges in Global South urbanizing contexts.

However, as well stressed by Pasquini and Enqvist (2019), "implementing and maintaining such projects within African cities has often taken a back seat to the provision of more urgent basic services such as housing, transportation and education facilities (Goodness & Anderson 2013; Gwedla & Shackleton 2015; du Toit et al. 2018)". This happens for several reasons:

- It is difficult to orient political decision-making between urgency (short term) and importance (long term), both for evident contingencies and for cultural motivations.
- Drafting the governance tools and policies to implement and maintain green infrastructure in an effective way is problematic in contexts where relevant stakeholders have been trained according to a completely different planning paradigm and prefer to adopt conventional approaches, especially in terms of vertical and horizontal institutional coordination.
- The benefits of green infrastructure are not clearly understandable and measurable for citizens and for most of the actors involved in decision-making.
- Specific data visualization and management tools are required to deal with the spatialization of certain information regarding WEF cycles, and skilled human resources and trans-disciplinary attitude can be scarce.
- Territorial scenarios promoting the development of green infrastructure might seem too abstract to be understood and implemented, being far from the potential action of small-sized stakeholders.

The case study of "Boa_Ma_Nhã, Maputo!", working on strategic planning scenarios for the Maputo Province, embraced the challenge to investigate the potential of a territorial armature approach, also dealing with the criticalities listed above, trying to cope with them, albeit within the limits of the project.

3   Strategic Territorial Scenarios for the Maputo Province: Towards an Integrated Metropolitan Vision

Maputo metropolitan system extends from the Indian Ocean coast (east) to Lebombo Mountains and the national border (west) and from the Maputo River (south) to the Incomati River (north), rapidly urbanizing along the main infrastructural corridors connecting Maputo to the secondary cities in the province. Land and water competition, food insecurity and energy scarcity, under the threat of climate change and in lack of any regional/metropolitan strategic planning tool, complete the critical framework.

The integration between planning indications from the municipalities and the districts of the Maputo Province and information collected and processed during the assessment phase of "Boa_Ma_Nhã, Maputo!" project (as described in Chaps. 4 and 5) helped in understanding current urbanization trends and challenges in respect with urban-rural relations.

The three districts of Boane, Moamba and Namaacha and the municipalities of Boane and Namaacha, as seen in previous contributions, are facing a number of key challenges in terms of socio-economic and urban-rural development, conditioning the future of the whole area. The main issues observed and identified during the analytical study of the region—in continuity with previous research conducted by Jenkins

(2013), Andersen et al. (2015) and Macucule (2010, 2016)—can be synthesized as follows:

- Differentiated rapid demographic growth associated with continuing high poverty levels and low access to basic educational, social and public health services.
- Growing environmental issues related to water and energy management and climate change in a time of increasing scarcity of basic resources, along with food (in)security, deforestation, soil consumption and impermeabilization in environmentally sensitive areas.
- A very low degree of interconnectivity and accessibility at the provincial and district scale between small and remote rural settlements and major urban areas
- Weak local government, lack of institutional and legislative capacity across municipal and district borders with the relative lack of state capacity to develop urban areas formally and to implement environmentally efficient policies as well as the lack of technical staff to survey the ground and enforce the existing plans and regulations to the citizens (municipal policing).
- The co-existence of processes of fragmentation and dispersion of urban activities and the centralization of social amenities (health, education and recreational facilities) workplaces and cultural services in Maputo, with consequent asymmetry in the quality of housing and access to basic infrastructure and services between central and peri-urban areas.
- The lack of job opportunities, especially of the female, younger and less skilled workforce, resulting in under-employment, "informal" activities in the commerce sector and intensive migration flows towards Maputo, but even abroad towards South Africa and eSwatini.

At the same time, despite being neglected in terms of strategic planning actions, the three districts and two municipalities are providing crucial services for the survival and growth of Maputo in terms of water, energy and food provision, as well as important educational services and ecological assets.

"Boa_Ma_Nhã, Maputo!" project, in the metropolitan perspective described above, focused on the need to recognize and re-balance such interdependencies. In particular, the project suggested overcoming the existing unbalanced dynamics in the region by re-defining the current image of Maputo from a centripetal urban system growing along with its radial infrastructure towards alternative development concepts built on open networks sustained by the growth of small and intermediate cities across the province. This territorial perspective supports the idea of cooperation among urban centres with different sizes to grow together as an integrated system of urban and rural nodes that cohesively plan their future instead of competing with the capital or among themselves multiplying splintering patterns.

Starting from early on-site observations and preliminary conceptualizations of potential territorial armatures to be strengthened, the main outcomes of "Boa_Ma_Nhã, Maputo!" project aimed at stressing the potential polycentric development of the Maputo Province to support alternative trajectories of urban-rural linkages and integrated approaches to growth through specific nexus-oriented scenarios and guidelines (Fig. 1). The objectives of this effort include designing and triggering

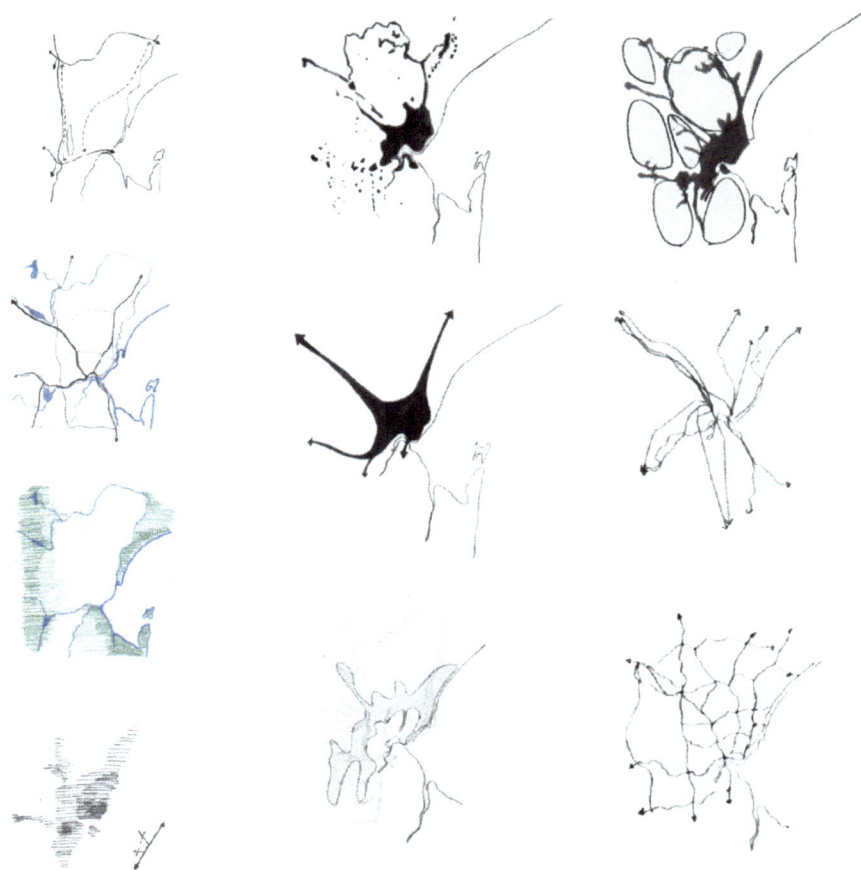

Fig. 1 Interpretative diagrams of Maputo's metropolization dynamics. *Source* Elaboration by "Boa_Ma_Nhã, Maputo!" (A. Frigerio), 2019–2020

the definition of territorial and trans-scalar infrastructures and planning/governance tools for socio-economic development with a focus on the integration between green infrastructure for water and food security with landscape and public space policies and projects (Montedoro et al. 2020).

In particular, building on the results of the assessment phase of the project, inter-weaving the results from the trans-disciplinary cartographic and analytical tools and from the activities in the field (Boa_Ma_Nhã, Maputo! 2020a, b), three sets of guide-lines have been drafted in relation to the food, water and energy systems, matching challenges and opportunities and investigating the spatial effects of the different recommendations in thematic areas of intervention (see Fig. 2). The drafting of

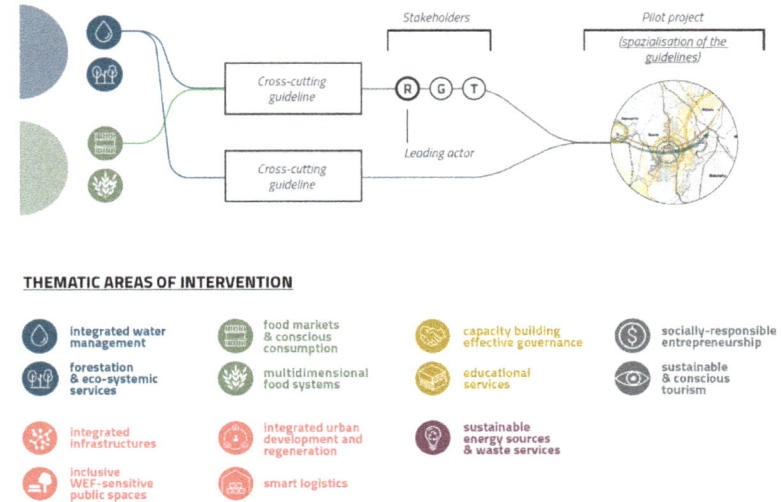

Fig. 2 Workflow, telescopic approach and thematic areas of intervention *Source* Elaboration by "Boa_Ma_Nhã, Maputo!" (A. Frigerio, A. Buoli) 2019–2020

guidelines moved the research focus from "problem setting" to "strategic thinking", opening to operative interventions as both physical transformations (i.e. reforestation initiatives/water-harvesting projects) and policy-oriented or research-led guidelines (i.e. green energy policies), always clarifying the potential actors involved (public, private, third sector). Subsequently, in a WEF nexus perspective, cross-cutting guidelines with effects on multiple thematic issues have been highlighted and spatialized in four territorial pilot projects. These have been identified in the shape of strategic but robustly spatially rooted scenarios (Fig. 3), to be intended as spatial images of potential strategic frameworks of cooperation among the involved districts, municipalities and stakeholders (public, private, third sector and research sector):

(1–2)   The Maputo-Boane-Namaacha Transect with a focus on the Namaacha food and tourism system (a, b).
(3)   The Maputo metropolitan agriculture green belt.
(4)   The Boane-Moamba ecotonal armature.

The identification of the pilot projects reflects the priority for "Boa_Ma_Nhã, Maputo!" to recommend investing in the identification, protection and valorization of continuous metropolitan blue and green infrastructure, robust armatures to safeguard and enhance the production of the essential ecosystem services for the survival and sustainable development of Maputo Province.

Due to climate change and socio-economic factors, vast areas of this territory are now non-productive landscapes, with a low value in respect to the potential production of ecological or agricultural services. The proposed scenarios promote landscape infrastructure as triggers for environmental and socio-economic processes

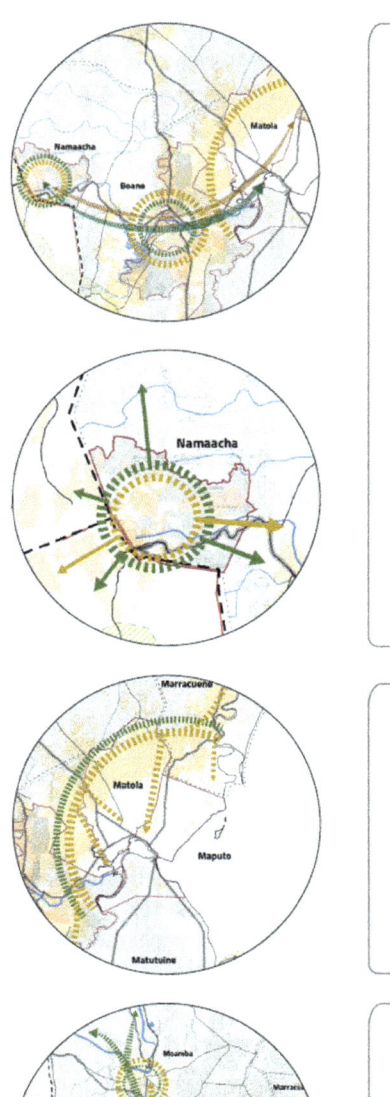

**1a | WEF-SENSITIVE SCENARIOS FOR
THE MAPUTO-BOANE-NAMAACHA TRANSECT**

- *A network of socio-ecological armatures supporting the sustainable growth of the region*
- *Food System: Re-orienting production, logistics, education and consumption*
- *Protected areas and energy production: from coal-based systems to renewable off-grid synergies*

1b | NAMAACHA FOOD & TOURISM SYSTEMS

- *An agro-ecological food system with high-quality small scale productions*
- *An eco-tourism system building on existing assets and making Namaacha a religious and cultural destination*
- *Regeneration of built heritage and public space*
- *Strengthening education and research offers*
- *Logistic facilities and innovative tools for trade and promotion*

**2 | MAPUTO METROPOLITAN AGRICULTURE &
GREEN BELT & IN-BETWEEN URBAN POCKETS**

- *Envisioning of the Maputo green belt as rurban metropolitan system balancing urbanization and agriculture economies*
- *Guidelines for a sustainable urban expansion with eco-district neighbourhoods*
- *Sustaining urban agriculture and local entrepreneurship for small-scale farming effectiveness and food security*

3 | BOANE-MOAMBA ARMATURE

- *A liner agro-armature directly connecting Boane to Moamba and the Maputo corridor, strengthening the potential role of Boane as food industry and trade hub*
- *Improvement of the road network*
- *Ecological protection of the beautiful landscape of the highlands, with promotion of eco-tourism activities*
- *Regulation of mining and extraction activities*

Fig. 3 Strategic pilot projects for the Maputo province *Source* Elaboration by "Boa_Ma_Nhã, Maputo!" (A. Frigerio, A. Buoli) 2019–2020

to counteract this tendency and generate resources, and act as civic instigators for new forms of stewardship and entrepreneurship. They are drafted by discussing actions promoted by the planning tools actually in force in respect to a WEF nexus perspective and combining them in a trans-scalar continuous system rooted in the territorial geographic structure of the area.

This spatial configuration is meant to be enriched by a layer of active or potential resources and actions that reveal the effective potential of highlighted green and blue systems to behave as a dynamic socio-ecological infrastructure through active processes of care, protection and sustainable exploitation. A set of examples of actions at the different scales is provided to foster imageability and engagement of potential funders and investors. Moreover, existing ongoing projects are considered as relevant processes to be set in synergy within a larger comprehensive strategy.

Pilot projects' scenarios should catalyse a wide range of public, private and non-profit actions with different scales, in support of commonly defined goals, and thus facilitate the development of the specific WEF-sensitive soft-governance tools (such as multi-sectoral public agencies, committees and consortia, agendas or road maps for decision-making) to be promoted and implemented to overcome existing administrative rigidities and to give socio-economic robustness to the sustainable development of the Maputo Province.

4   The MaBoNa Transect: A Telescopic Scenario for an Economic, Ecological and Cultural Territorial Armature

The Maputo-Boane-Namaacha (MaBoNa) Transect, one of the pilot projects proposed by "Boa_Ma_Nhã, Maputo!", has been a test bed for a potentially replicable methodology to envision a telescopic and integrated territorial scenario.

MaBoNa Transect, structured by EN2 and the Umbeluzi and Impaputo Rivers, is one of the most important metropolitan infrastructural systems in the Maputo Province: a historical, economic, cultural and ecological resource of extraordinary value for the whole metropolitan system. The transect, spanning for barely 80 km perpendicular to the coast and covering an altitude difference of around 600 m, crosses several different landscapes with rich topography and geology, such as urbanized areas, suburban informal "sprawl", fertile agricultural lands, beautiful natural landscapes, small villages and industrial excavation sites (Fig. 4).

Respecting this diversity of characters and identities, the MaBoNa Transect project proposes a reconsideration of the territorial relationships with the river and the road, suggesting a shared comprehensive vision for the future of this rich territory: a linear network of green, blue and eco-urban systems promoting the protection and sustainable use of local natural resources.

The river has shaped the landscape and different cultures, uses and ecosystems have gathered around its stream. During the twentieth century, the relationships of

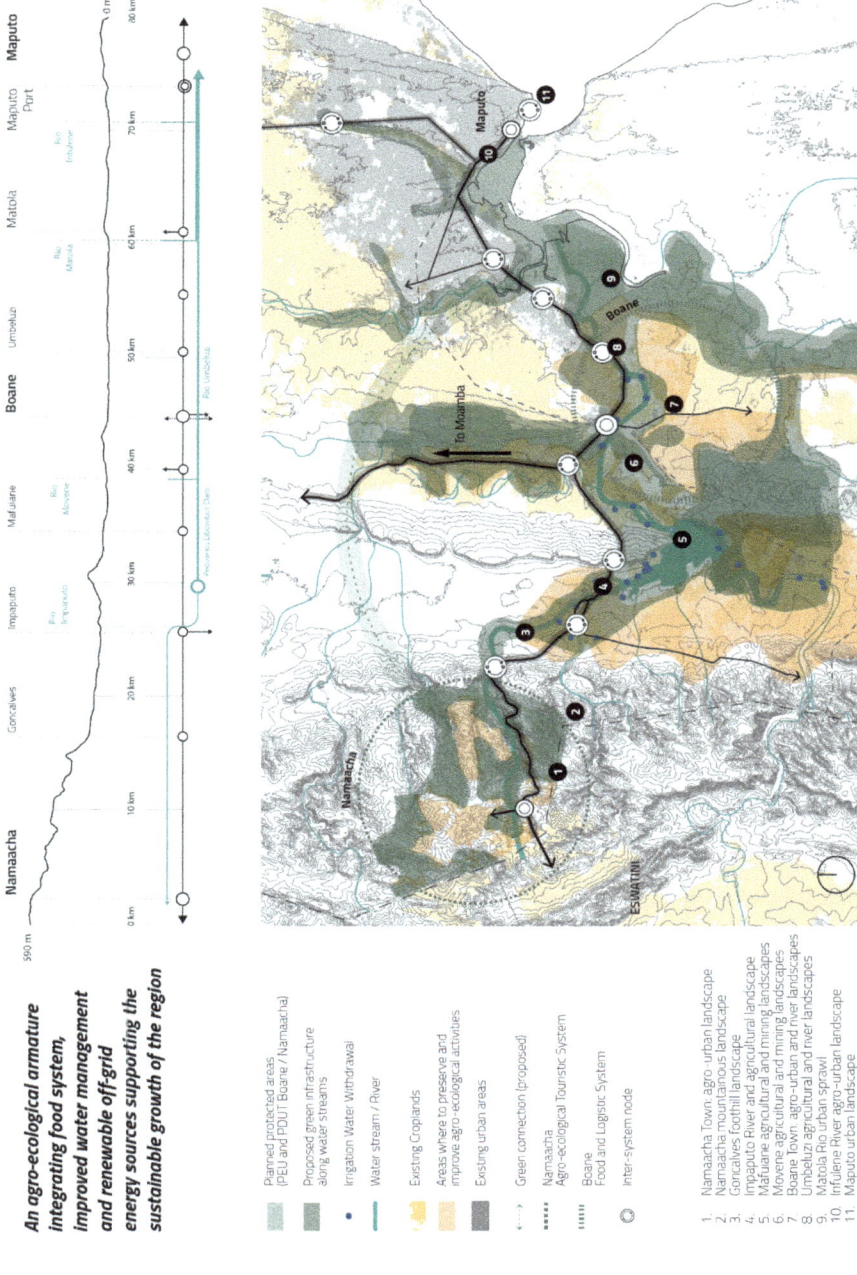

Fig. 4 Strategic WEF-sensitive scenario for the Maputo-Boane-Namaacha transect Elaboration by "Boa_Ma_Nhã, Maputo!" team 2020 (A. Buoli, A. Frigerio, A. Macchiavello)

the Umbeluzi's watershed area with Maputo (formerly Lourenço Marques) became more and more important, as it became an essential provider of services to the city. At the beginning of the century, the Umbeluzi River was the primary source for the urban water system, and, in the fertile downstream Boane area, experimental farms started modern agricultural research and production of food and cash crops. At the same time, Namaacha, upstream, thanks to its mild climate, became a destination for tourism, quality education, mineral water bottling and experimental forestation projects.

Since 1987, with the Pequenos Libombos Dam and reservoir becoming operational, the ties between the MaBoNa Transect and the city have become even stronger. However, more recent changes in transboundary relationship with eSwatini and the socio-economic context's evolution have led to the need of reconsidering and re-empowering metropolitan rural-urban connections.

The current challenges related to the water, food and energy systems at the metropolitan scale, in fact, partially depend on the impoverished performances of the MaBoNa landscape infrastructure, combined with the scarcity of investments and the limited capacity of this system to adapt to climate change and contemporary needs, despite its high potential.

The MaBoNa Transect is meant to be a generational project to be started by an engaging strategy around brain-storming scenarios, with the aim of consolidating a clear identity to this territorial armature and building an inspiring vision, a plan for the river and the road corridor as metropolitan and regional landscape infrastructures, while showcasing their natural beauty. The goal of the project should be to provide answers to current issues, as well as predicting future growth and challenges, and attempting to re-balance the rural-urban metabolism on a long-term perspective.

The first step to trigger this process was to investigate two complementary strategies related to ecological and food systems, together with a set of sample actions focused on the different nodes (or hubs) of the transect.

The diagram of the agro-ecological system (Fig. 5, upper part) represents a continuous green and blue corridor supporting environmental protection and the production of ecosystem services by synthesizing and synergistically recombining the existing fragmented mosaic of protected areas and ecological corridors included in the regulations and planning tools in force. The protection of the Umbeluzi and Impaputo Rivers corridor will improve the ecological health of the water basin, improving water quality, quantity and management to re-balance water competition.

Moreover, re-forestation, ecological restoration, biodiversity enhancement, heritage protection, eco-tourism and public space promotion, together with the definition of specific guidelines to harmonize urbanization and urban regeneration, will help to address the challenges of carbon sequestration, climate adaptation and will promote social and cultural uses, awareness and care. In the Boane area, the priority is the conservation of natural assets threatened by uncontrolled urbanization, while in Namaacha, the most relevant opportunity is in the restoration of impoverished ecosystems, also through industrial and community re-forestation initiatives.

The diagram of the food system (Fig. 5, lower part) is a first attempt to spatialize the correlation between each stage of the food cycle (production, logistics, selling,

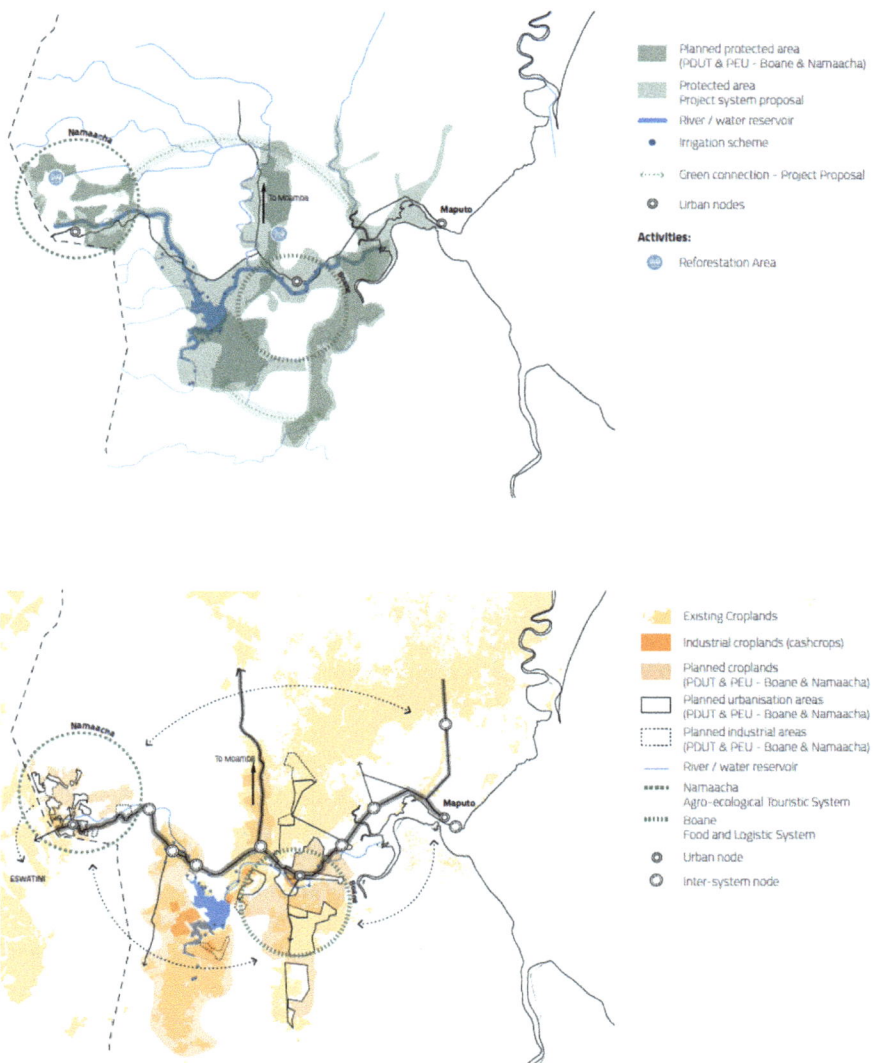

Fig. 5 Agro-ecological and food systems for the Maputo-Boane-Namaacha transect Elaboration by "Boa_Ma_Nhã, Maputo!" team 2020 (A. Buoli, A. Frigerio, A. Macchiavello)

transformation, consumption and waste management) along the potential MaBoNa agro-metropolitan axis. The EN2 has the potential to become a structural armature for the food system, with logistic and trade nodes located in strategic points all along its path, to facilitate the development of the surrounding countryside and promote access to the market and transformation of local products.

The scenario impacts all the agricultural typologies, from cash-crop industries to agricultural cooperatives, subsistence farming or urban agriculture, all required to contribute in synergy to regional food security. It also tackles the current designation of vast agricultural areas in Boane and Namaacha to future urbanization by suggesting alternative settlement models inclusive of small-scale farming. At the same time, it promotes the exploitation of currently non-productive lands through innovative irrigation and farming systems. Technical education and research, in line with the history of the area and capitalizing on the existing agrarian schools along the transect, are also credited as crucial assets of the food system. The scenario reveals the potential role of Boane as an industrial-agriculture and transformation hub, as well as a redistribution centre for agricultural products at the regional and international levels, and the potential vocation of Namaacha as a small-scale, high-quality agroecological research and farming hub.

Existing ongoing projects of any scale that have been considered relevant in contributing to the larger comprehensive strategy have been mapped and included in the scenario as crucial triggers for initiating and facilitating the process of recognition, investment and implementation.

Moreover, a series of sample actions have been designed to clarify the potential spatial implementations of the cross-thematic guidelines. Sample actions represent potential triggers of physical transformation at various scales and in relation to different spatial dimensions: from the macro-territorial level of the Umbeluzi Water Basin and the Maputo Watershed (blue and green) infrastructures to the local scale of neighbourhood regeneration and sustainable micro-urban planning as eco-districts, to public indoor and open spaces, and alternative energy-production systems (Fig. 6). Each action results from a combination of different WEF issues, fields of interest, collaborations among consortia of stakeholders and funding formats. Actions also combine a sequence of strategies or recommendations to be carried out to perform and achieve the main aims of the action itself. A list of references is also provided, with projects or policies already implemented serving as benchmarks for the development of the action. Each action is accompanied by some iconographic materials, considering visualization extremely relevant for communicating and circulating ideas and concepts.

Sample actions are also relevant to better clarify the telescopic nature of the scenario, which could be zoomed in at any point, revealing a constellation of ongoing initiatives, agencies and future potential projects that contribute to making implementation possible and amplifying the system's resiliency.

"Boa_Ma_Nhã, Maputo!" dedicated a specific focus to explore the telescopic nature of the scenario by zooming in on one of the nodes of the MaBoNa Transect: the Namaacha node (Fig. 7).

The study of the available documents and planning tools, the production of analytic cartography and the research conducted on-site in interaction with local authorities and stakeholders, allowed us to unfold the main challenges and potentialities for social, economic and territorial development of Namaacha, which regard specifically its agro-ecological and food systems, as well as its built heritage, building on

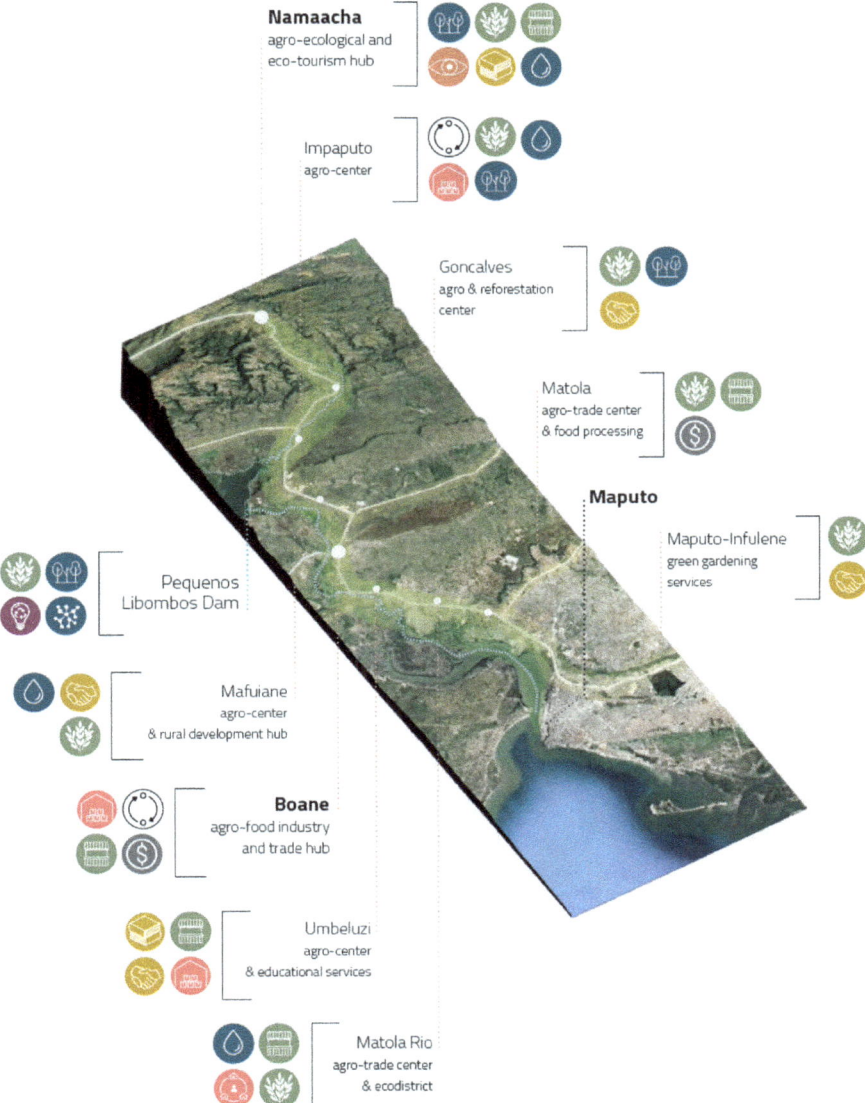

Fig. 6 A tool kit of actions for the Maputo-Boane-Namaacha transect Elaboration by "Boa_Ma_Nhã, Maputo!" team 2020 (A. Buoli, A. Frigerio)

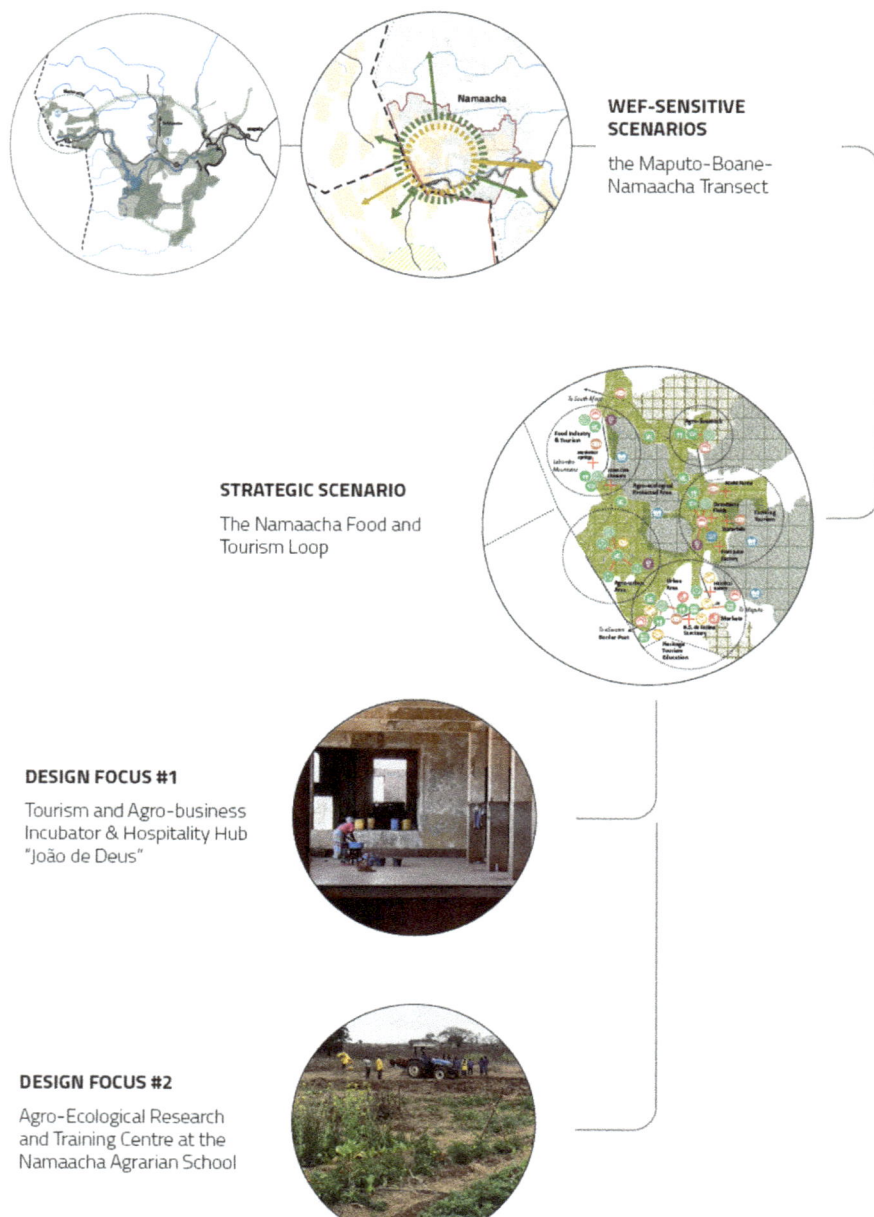

WEF-SENSITIVE
SCENARIOS

the Maputo-Boane-
Namaacha Transect

STRATEGIC SCENARIO

The Namaacha Food and
Tourism Loop

DESIGN FOCUS #1

Tourism and Agro-business
Incubator & Hospitality Hub
"João de Deus"

DESIGN FOCUS #2

Agro-Ecological Research
and Training Centre at the
Namaacha Agrarian School

Fig. 7 Local development plan for Namaacha Elaboration by "Boa_Ma_Nhã, Maputo!" team 2020
(A. Buoli, A. Frigerio, A. Macchiavello)

its vocation as a key destination for religious, cultural and eco-tourism. Reconceptualizing and setting in synergy these value assets as resiliency drivers, the project aimed at leveraging local development by proposing a locally framed strategy and governance proposal coherent with the regional one and including two pilot projects whose feasibility has been discussed with local stakeholders for a real potential implementation.

Thus, MaBoNa Transect has been telescopically explored, in design terms, from the regional strategic vision to a potential local implementation project, as a sample of one of the many that international cooperation and local entrepreneurship are regularly proposing and realizing, but out of any synergic framework.

5   Closing Remarks. The Relevance of Strategic Spatial Scenarios as Common Grounds for Cooperation

Through a systematic collection and multimedia narrative—drawn on the combined use of maps, diagrams, scenarios and sample actions—"Boa_Ma_Nhã, Maputo!" research project proposed a series of ideas, strategies and images for the Maputo Province, introducing proposals for alternative planning tools, stressing a transdisciplinary and trans-scalar WEF-sensitive agenda and opening the dialogue on the future of these territories.

The aim was, in the first place, to provide strategic spatial scenarios as interpretative and imaginative framework for supporting further research trajectories and policy-making initiatives by local authorities and the urban governance stakeholders in the Boane, Moamba and Namaacha area as well as for the entire Maputo metropolitan region.

The emphasis is not only on the normative tools or the formal planning instruments that are currently in force—that served as a crucial knowledge base—but also on a variety of different "gazes" that might trigger new strategic and operative visions (at different scales) that have been little explored by current plans and policies thus far.

From territorial guidelines to local development plans and projects, drafting strategic scenarios could mean co-producing platforms for building knowledge, promoting negotiation and facilitating decision-making. Designing them as telescopic tools could match synthetic strategic thinking and pragmatic implementation tools, contextualizing know-hows and integrating complementary approaches.

The guidelines, scenarios and actions here proposed are drawn on two general assumptions: on the one hand, the need for more active multi-level dialogues between the actors of the local governance; on the other hand, the urgency to connect and integrate the different strategies and visions that have been planned and adopted by the different administrative entities, in view of the current urbanization and demographic trends in the area.

The scenario for the MaBoNa Transect aims, in fact, to "sew together"—in terms of agroecological and mobility infrastructures—and to "fill in" the gaps between the strategies proposed by the plans for Boane and Namaacha (both at the level of the PDUT and PEU). At the same time, it also suggests a broader vision for this crucial section of the Maputo Province, recalling some current and previous strategic proposals for the area (for instance, the 1969 Urbanisation Master Plan of Lourenço Marques).

Finally, the collection of ideas and actions proposed by the project reports (Boa_Ma_Nhã, Maputo! 2020c) delivered to the various institutional actors involved should be seen as an attempt to integrate micro-scale/local initiatives into broader trans-scalar visions and strategies in the different "pilot areas" identified by the plan. This effort suggests that the long-term effectiveness and sustainability of local development projects can only be achieved if sustained and strengthened by solid governance and administrative frameworks that at the same time should be flexible enough to adjust to contingencies and sudden changes (the global COVID-19 crisis is just an example).

This seems to be necessary to avoid the mismatch between very specific and tailored actions (often with a strong local commitment and participation) and "the bigger picture", namely, national or provincial governmental programmes. Such criticism is also part of the local planning debates and is shared by colleagues, experts and policy-makers operating in the Maputo Province and consulted along the whole research process.

In other words, the coordination between the various institutional stakeholders should be strengthened both horizontally—thus avoiding sectorialization—and vertically, between the various levels of government, a challenge that also engages the countries of the global North in planning for sustainable urban development.

These tasks, as clearly emerged by the "Boa_Ma_Nhã, Maputo!" activities in the field, are anything but easy, especially considering the previously mentioned mismatch in terms of priority between short-term urgency and long-term importance. Thus, the first objective of research projects like this should always be circulating ideas, raising awareness, engaging local actors of any kind and clearly showing feasible alternative ways of "doing things differently" by building inclusive common grounds to dialogue, share and envision together the future.

References

Abbott J (2012) Green infrastructure for sustainable urban development in Africa. Taylor and Francis

Andersen JE, Jenkins P, Nielsen M (2015) Who plans the African city? A case study of Maputo: part 1—the structural context. Int Dev Plan Rev 37(3):331–352

Behrens R, Watson V (1996) Making urban places. Principles and Guidelines for Layout Planning. UCT Press, Cape Town

Boa_Ma_Nhã, Maputo! (2020a) Assessment Report. Politecnico di Milano, Milan

Boa_Ma_Nhã, Maputo! (2020b) Planning Tools Report. Politecnico di Milano, Milan

Boa_Ma_Nhã, Maputo! (2020c) Polisocial Development Plan. Politecnico di Milano, Milan

Chiesa A (2010) Misura e scala della città contemporanea. Il caso barese. Landscape Urbanism e Morfologia Urbana: il disegno urbano italiano verso il disegno urbano anglosassone. Dissertation. DiAP–DrPAU, Politecnico di Milano

Dematteis G (1997) Le città come nodi di reti: la transizione urbana in una prospettiva spaziale. In: Dematteis G, Bonavero P (eds) Il sistema urbano italiano nello spazio unificato europeo. Il Mulino, Bologna

du Toit MJ, Cilliers SS, Dallimer M, Goddard M, Guenat S, Cornelius SF (2018) Urban green infrastructure and ecosystem services in sub-Saharan Africa. Landscape and Urban Planning 180

EC—European Commission (2015) Towards an EU Research and Innovation policy agenda for Nature-Based Solutions & Re-Naturing Cities. Final Report of the Horizon 2020 Expert Group on 'Nature-Based Solutions and Re-Naturing Cities'. European Union, Luxembourg. Accessible via EC. https://ec.europa.eu/programmes/horizon2020/en/news/towards-eu-research-and-innovation-policy-agenda-nature-based-solutions-re-naturing-cities. Accessed 12 December 2021

Egoz S, Makhzoumi J, Pungetti G (2011) The right to landscape. Contesting landscape and human rights. Routledge, London and New York

Goodness J, Anderson PM (2013) Local assessment of Cape Town: navigating the management complexities of urbanization, biodiversity, and ecosystem services in the Cape Floristic Region. In: Elmqvist T, Fragkias M, Goodness J, Güneralp B, Marcotullio PJ, McDonald RI, Parnell S, Schewenius M, Sendstad M, Seto KC, Wilkinson C (eds) Urbanization, biodiversity and ecosystem services: challenges and opportunities. Springer, Dordrecht, pp 461–484

Gouverneur D (2014) Planning and design for future informal settlements: shaping the self-constructed city. Routledge, New York

Gwedla N, Shackleton CM (2015) The development visions and attitudes towards urban forestry of officials responsible for greening in South African towns. Land Use Policy 42:17–26

Harrison P, Bobbins K, Culwick C, Humby T, La Mantia C, Todes A, Weakley D (2014) Urban resilient thinking for municipalities. University of the Witwatersrand, Johannesburg

Jenkins P (2013) Urbanization, urbanism, and urbanity in an African city: home spaces and house cultures. Springer, Cham

Lefebvre H (1968) Le droit a la ville. Anthropos, Paris

Macucule D (2010) Metropolização e Restruturação Urbana: O território da Grande Maputo. Dissertation, New University of Lisbon. Department of Geography and Regional Planning

Macucule D (2016) Processo-forma urbana: Restruturação urbana e governança no Grande Maputo. Dissertation, New University of Lisbon. Department of Geography and Regional Planning

Montedoro L, Buoli A, Frigerio A (2020) Towards a metropolitan vision for the Maputo province. Maggioli Editore, Santarcangelo di Romagna.

Ortiz P (2014) The art of shaping the metropolis. McGraw Hill

Pasquini L, Enqvist JP (2019) Green Infrastructure in South African cities. Report for Cities Support Programme Undertaken by African Centre for Cities, May 2019. http://www.africancentreforcities.net/wp-content/uploads/2020/01/CSP_green-infrastructure_paper_LPasquini_JEnqvist_11.pdf. Accessed 12 December 2021

Prosperi D, Moudon A, Claessens F (2009) The question of Metropolitan Form: an introduction. In: Claessens F, Vernez Moudon A(eds) Metropolitan Form. Footprint Volume 5, Autumn 2009, TU Delft

Ramaswami A, Russell AG, Culligan PJ, Rahul Sharma K, Kumar E (2016) Meta-principles for developing smart, sustainable, and healthy cities. Science 352(6288):940–943

Roe M, Mell I (2012) Negotiating value and priorities: evaluating the demands of green infrastructure development. J Environ Plann Manag 1–12

Secchi B (2000) Prima lezione di urbanistica. Laterza, Bari

Shane D (2005) Recombinant Urbanism. John Wiley & Sons, New York

Swilling M (2013) Reconceptualising urbanism, ecology and networked infrastructure. In: Pieterse E, Simone AM(eds) Rogue urbanism. Jacana Media and the Centre for African Studies, University of Cape Town, Auckland Park, South Africa

Part III
A Transdisciplinary Lexicon of Sustainable Planning in Sub-Saharan Africa

Agriculture and Food Security: Implications on Sustainable Development and the WEF Nexus

Maria Cristina Rulli, Davide Chiarelli, Nikolas Galli, and Camilla Govoni

Abstract Agriculture is the backbone of the Mozambican food system, and it represents not only the most important economic sector in the country but also the main user of natural resources such as water, land and eco-systemic services. However, more than 95% of agricultural areas consists of mostly rainfed smallholder farms, characterized by low yields and, when irrigated, by low irrigation efficiency. The wide yield gap affecting most Mozambican cultivations is partly due to the unfertile sandy soil and to the prevalence of rainfed agriculture. In addition to that, flawed farming practices and scarce mechanic agricultural inputs contribute to the low productivity of farmlands in Mozambique, hindering a diffuse and adequate access to healthy and nutritious food for the local populations. In addition, the overall resilience of the agricultural system is low also regarding climate and weather extremes. The fragilities of the Mozambican agricultural system and their consequences on food security and malnourishment have led to a series of acts and policies aimed at the development of agriculture and the eradication of hunger. Based on a WEF Nexus perspective, the chapter presents the overall conditions of water and agriculture in Mozambique and discusses the main national policy framework to address and reduce food insecurity.

1 State of Water, Agriculture and Food Security in Mozambique

One of the best ways to understand how a country manages its water resources is to understand where water resources come from. The precipitation pattern in Mozambique is characterized by relevant variations across the country (Fig. 1). The rainy season starts in October and ends in April. The country average of annual

M. C. Rulli (✉) · D. Chiarelli · N. Galli · C. Govoni
Department of Civil Engineering (DICA), Politecnico di Milano, Milan, Italy
e-mail: mariacristina.rulli@polimi.it

D. Chiarelli
e-mail: davidedanilo.chiarelli@polimi.it

155

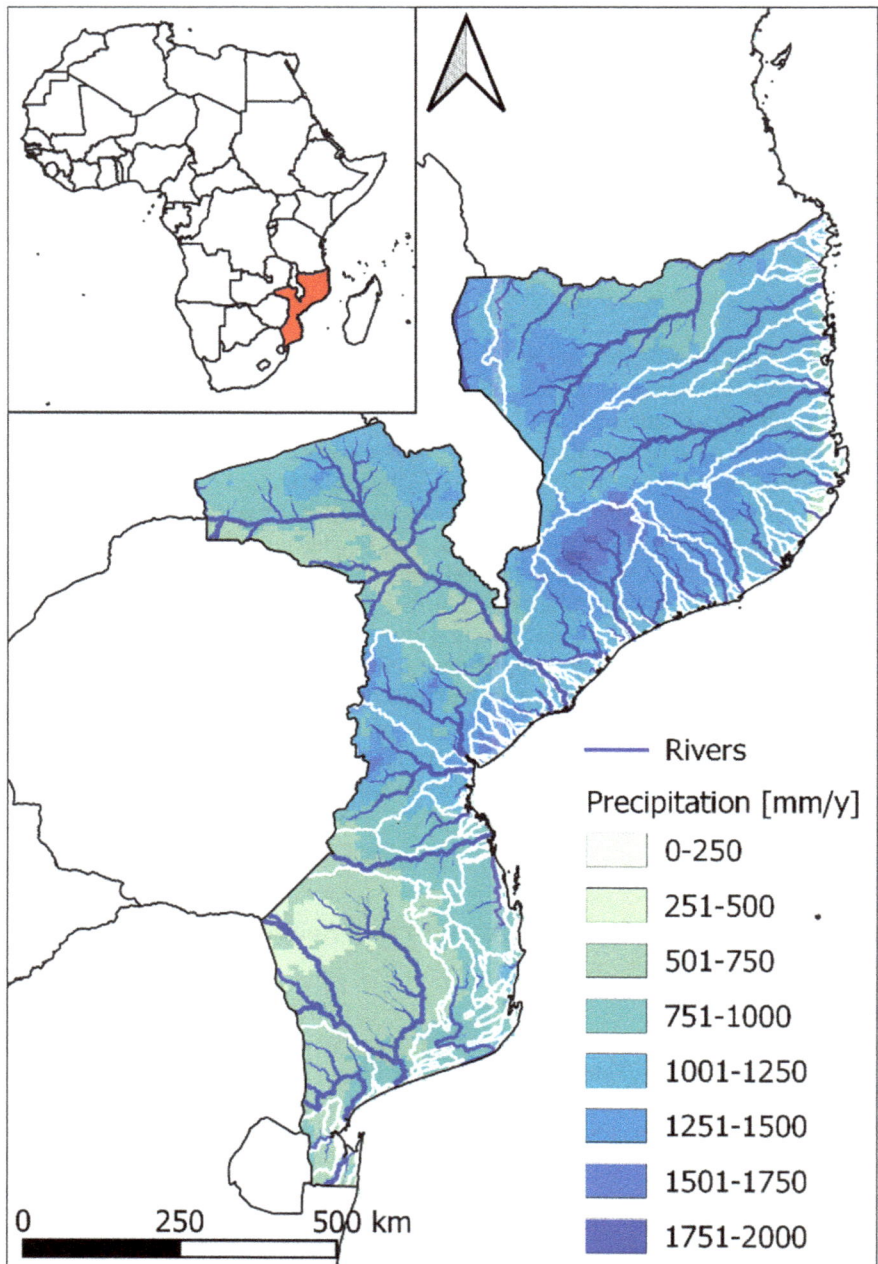

Fig. 1 Location, precipitation pattern and rivers of Mozambique. White borders are watershed limits, taken, as the rivers, from the hydrosheds database (Lehner et al. 2008), whereas precipitation is taken from CHIRPS (Funk et al. 2015)

precipitation is 1032mm, but it can reach 1200mm in the central coastal area (Funk et al. 2015). Other coastal regions have annual precipitation values ranging between 800mm and 1000mm. These values decrease to 400mm when moving towards the inland areas, except for the central-northern mountain regions, where the monsoon can deliver up to 2000mm of yearly precipitation. This precipitation contributes to the 100km3/y of internal renewable water resources. The total renewable water resources, including water entering the country from transboundary basins, amounts to 217km3/y, thus resulting in a 54% dependency ratio of Mozambique to external water resources (FAO 2016).

Of the 104 river basins in Mozambique, 9 are shared with neighbouring countries (Lehner et al. 2008) (Fig. 1). Some of these transboundary river basins play an important role for water supply and agriculture, both in Mozambique and in the other countries. For instance, the Limpopo River basin, shared with South Africa, Zimbabwe and Botswana, supplies water to the largest irrigation scheme in Mozambique, the Chókwè irrigation scheme (de Sousa et al. 2019; Ducrot et al. 2020). The Zambesi River basin accounts for 50% of the surface water resources and 80% of hydropower potential (FAO 2016). Being Mozambique a predominantly coastal country, it locates downstream in the transboundary basins, thus often being subjected to the decisions of upstream countries.

Water withdrawals in Mozambique amount to 0.884km3. Industry accounts for 2% of withdrawals and domestic use accounts for 25%. Drinking water supply in urban areas and rural settlements is withdrawn mainly from groundwater sources, whereas the remaining, and by far larger, part of withdrawals is from surface water sources (FAO 2016). Agriculture accounts for the remaining 73% of water withdrawals. Eighty-nine per cent of agricultural withdrawals is used for irrigation, corresponding to an average volume of 0.165km3 over the period 2011–2016 (Chiarelli et al. 2020), whereas the rest is used for livestock and forestry. This high percentage of irrigation withdrawals contrasts with a very low extent of effectively irrigated areas (62000ha), especially in relation to Mozambique's irrigation potential (3Mha). In fact, irrigation is performed more extensively by agricultural companies, which use mainly sprinkler irrigation systems with efficiency rates up to 70%. However, more than 95% of agricultural areas consists of mostly rainfed smallholder farms, characterized by low yields and, when irrigated, by low irrigation efficiency (25 ÷ 50% for surface irrigation). The wide yield gap affecting most Mozambican cultivations is partly due to the unfertile sandy soil and to the prevalence of rainfed agriculture. In addition to that, flawed farming practices (e.g. selection of more susceptible varieties) and scarce mechanic agricultural inputs contribute to the low productivity of farmlands in Mozambique (CARE 2017). The overall resilience of the agricultural system is low also in regard to climate and weather extremes. In fact, more than 50% of harvest in the south of the country is at risk of being lost due to droughts or floods (FAO 2016).

Smallholder agriculture is mainly subsistence farming. The main crops are staple crops such as rice, maize, sorghum and cassava, some of the main contributors to the local diet. Still, Mozambique is characterized by an agricultural trade deficit, as it imports more agricultural goods than it exports (FAOSTAT 2018). Self-sufficiency at the national level has been achieved for cassava and beans, while consumption exceeds production for wheat, all vegetable oils and animal source food (CARE 2017; Ferrão et al. 2018). Price fluctuations are strongly seasonal for crops that are more linked to the internal market (e.g. maize), since most agriculture is rainfed, and thus highly sensitive of precipitation patterns. Clearly, on the other hand, crops whose market is dominated by import are more influenced, in terms of price, by the global market (Donovan and Tostao 2010).

Issues such as the low productivity of the domestic agricultural sector and the dependence, for many important staple crops, on import, contribute to the food insecurity situation affecting the country, both in terms of availability and access (CARE 2017). Moreover, a diet based mostly on cereals and tubers poses constraints also on the utilization pillar of food security, and thus on nutrition security, as it limits the supply of micronutrient-rich foods. As a result, malnourishment is a common issue in Mozambique. Forty-four per cent of children under the age of 5 are chronically malnourished and 18% are underweight, a percentage that can rise to the double in rural areas (Grabowski et al. 2013; CARE 2017; Ferrão et al. 2018). On the other hand, cash crops grown by agricultural companies are destined partly to the internal market and partly to export. These crops include tobacco and cotton, but the main role is played by sugarcane. In fact, half of the irrigated areas in Mozambique are destined to sugarcane cultivation for bioethanol production (FAO 2016; Ferrão et al. 2018).

Mozambique has issued a series of policies, plans and norms to structure its water governance system, improve the agricultural sector and thus reduce food insecurity. These actions and their effects are analysed in the following section.

1.1 Policies for Water Governance, Agricultural Development and Food Security

The gap between formal water rights and their effective application is still wide in Mozambique (Veldwisch et al. 2013). The Article 98 of Mozambique's Constitution states that 'Natural resources […] in inland waters […] shall be the property of the State' (República De Moçambique 1991). The Water Act of 1991 transposes this article and states that internal waters, surface water bodies and their beds and subsurface water bodies are property of the State. The *Administraçoes Regionais de Águas* (ARA), i.e. the Government Water Boards, are established by the Water Act as public institutions aimed at managing watersheds. Their territorial scope coincides with one or more river basins. The Water Act also separates common use of water from private use. In particular, agricultural water use is defined as common use for field sizes up to 1ha without mechanical water extraction/derivation facilities. Farmers or companies irrigating more than 1ha must apply for a licence to their local ARA. This process is disciplined by the Decree 43/2007 - National Water Policy (República De Moçambique 2007a). The applicant must declare the amount of water and its source, and the ARA must decide whether to grant the licence, depending on the availability of the requested water, upon payment of a fee. Water is assigned among different dwellers following a 'first come, first served' principle. Although clearly aimed at water conservation and resource monitoring, this aspect of the Water Law produces some negative side effects. First, the issue of determining water availability is not a trivial one, not only because of the difficulties inherent in the task itself. In fact, common use of water is prioritized over private use. Thus, knowledge of current common use should be crucial to determine the amount of water available for private use. However, this knowledge is not achieved for most of the cases. Therefore, the priority is given *de jure* to an unquantified common use of water, which remains *de facto* ignored when planning the granting of licences for private use. Second, the limit of 1 ha and no mechanic equipment for agricultural common water use may pose constraints to the development of small-scale agriculture. For instance, whenever a smallholder decides to improve its farm, either in terms of size or of mechanic inputs, their water use automatically moves from common to private. The subsequent need of a formal license, and the associated fee, often put smallholders in front of the choice between giving up the improvement or acting outside the law (Veldwisch et al. 2013).

The National Irrigation Strategy of 2010 (Ministério da Agricultura 2010), the National Water Policy of 2016 (Decree 42/2016, República De Moçambique 2016a) and the Decree 43/2016 (República De Moçambique 2016b), furtherly deepening the irrigation aspects of the antecedent decree, all aim to provide better regulation to the water use in the agricultural sector.

The National Irrigation Strategy poses ambitious short-, medium- and long-term agricultural development targets, concerning agricultural and livestock productivity, but also irrigation expansion and the operationalization of irrigation infrastructures and the agricultural sector in general.

In response to this strategy, the *Instituto Nacional de Irrigaçao* (National Institute of Irrigation) has been established as an autonomous institution in 2012, with the aims to collect, systematize and share information about existing irrigation facilities, prioritize those having a higher development potential and streamline the technological improvement, as well as the expansion, of such facilities. Another important aim of the *Instituto Nacional de Irrigaçao* has been the development and implementation of the national irrigation plan (República De Moçambique 2016b) as it was anticipated in the National Irrigation Strategy.

The National Water Policy translates into norms and prescription all the water aspects of the Sustainable Development Goals, proposing specific practices for urban, peri-urban and rural areas. To this aim, it includes, for instance, objective quantifications of minimum service requirements for common water use, thus aiming at filling one of the gaps of the Water Law. Moreover, it defines interventions for promoting water safety and water recycling, such as good practices in recollection, transport, storage and conservation of domestic water. More in general, the National Water Policy defines not only objectives and policies regarding water use in all productive sectors but also its social and environmental reflections, with a special focus on environmental risk connected to water, typically represented, in Mozambique, by floods and droughts.

The fragilities of the Mozambican agricultural system and their consequences on food security and malnourishment have led to a series of acts and policies aimed at the development of agriculture and the eradication of hunger. The Food and Nutrition Security Strategy (ESAN) has been approved in 1998, revised in 2007 (ESAN II) and is currently being updated for a third version (ESAN III) (República De Moçambique 2007b). Its main objectives are to safeguard the country's food self-sufficiency, decrease the malnutrition incidence, also by improving the farmers' resilience to weather and climate shock, and to increase food availability and access in rural areas. In fact, one of the main innovations of ESAN II with respect to ESAN is the explicit definition of food access and nutritious food as a human right and as a responsibility of the state. The National Multisectoral Action Plan for the Reduction of Chronic Malnutrition (PAMRDC) (República De Moçambique 2010) aims at reducing the chronic malnutrition among children below 5 years of age, from the 44% value of 2008 to 30% in 2015 and 20% in 2020. Moreover, it defines a set of objectives for the strengthening of the food system, with a special focus on the weaker sections of society, e.g. adolescents, children and pregnant or lactating women. For each objective, it defines actions that have already been completed and new intervention targets at the household level, concerning, for instance, nutrition education, controlling infections and anaemia, improving food storage and processing and salt iodification. These strategies, plans and interventions, directly aimed at the achievement of food security, require specific measures to foster the development of the agricultural sector for it to be able to support the fight against hunger. In this regard, the Strategic Plan for the Development of the Agricultural Sector for 2011-2020 (PEDSA 2011) (República De Moçambique 2011) seeks to transform the predominantly subsistence-based agricultural sector into a more competitive one. It articulates its objectives into four pillars: increase agricultural productivity and competivity while achieving a healthy diet, increase access to market, use water, land, forestry and wildlife sustainably, create stronger agricultural institutions. In response to these pillars, the National Investment Plan for the Agricultural Sector for 2013–2017 (PNISA) (Ministério da Agricultura 2012) aims at operationalizing them while reaching a + 7% growth of the agricultural sector yearly, reducing by half the population suffering from hunger, and respecting the malnutrition reduction goals of the PAMRDC. However, this plan, not unlike others, does not constitute a real enforcement measure of the national strategies and objectives, as it covers multiple areas of action without getting into detail on any of them, consequently lacking both a priority scale and awareness about the necessary and available resources.

2 Implications on Sustainable Development and WEFE Nexus

Agriculture is at the root of the food system, and it represents not only the most important sector in Mozambican economy but also the main user of natural resources such as water, land and ecosystem services. Therefore, it is important to consider it in the context on one hand of sustainable development (in its social, environmental and economic components) and, on the other hand, of the Water Energy Food Ecosystem Nexus.

Mozambique's recent history is often regarded as a success story: although still ranking low on the main social and economic indicators, its growth after the civil war has been considerable (The World Bank Group 2019). However, it remains a country with several social issues, due to its structural constraints but also to the strong patriarchal culture. Indeed, as it is often the case, gender issues are deeply intertwined with the theme of agricultural development. Mozambican women are strongly involved in agriculture, as they constitute 65% of its labour force, yet they are excluded from planning and decision-making (Pellizzoli et al. 2010). Besides women, also migrant groups and lower social classes have historically been excluded from access to resources in favour of men from richer classes and more settled groups (FAO 2016; CARE 2017). Including these categories, especially women, in the decision processes would be beneficial not only in terms of gender equality but also because these excluded social groups are the ones most suffering from food insecurity. Thus, such an inclusion process would be in line with the aims of many of the policies mentioned above.

The contradictions between what is defined by agricultural, food security and water management policies, and what is effectively put into practice regard not only social development but also economic growth and the environmental context. Although many Mozambican rivers, especially the transboundary ones, are under severe pressure from different users, they are much better off from the viewpoint of pollution. In fact, given the low use of fertilizers by smallholders, nitrogen and phosphorous contamination in water bodies is often negligible (FAO 2016). However, some environmental risk arises from uncontrolled sewage of wastewater from industries and major urban centres. Moreover, artisanal mining and the transformation of forests into agricultural lands increase the risk of riverbed erosion and sedimentation (FAO 2016).

The many lakes and the major rivers flowing through Mozambique provide potential not only for irrigation but also for hydroelectric energy production. Hundreds of small dams providing collectively 60 Mm3 of water, 90% of which for irrigation, have supposedly been destroyed during the independence war (FAO 2016). On a larger scale, several dams exist on different rivers to provide Maputo with freshwater. The Cahora Bassa Dam on the Zambesi River, with 2060MW of installed capacity, is the largest hydroelectric plant in southern Africa (FAO 2016; Nielsen et al. 2016). It was initially built as an irrigation and flood prevention infrastructure. However, the periodical floods it prevented had a relevant fertilizing function, as they released nutrient-rich sediments on an area with a strongly irregular rain pattern. Moreover, the dam altered the inflow to the Zambesi Delta area, with remarkable effects on water security and on the delta ecosystem (Nielsen et al. 2016).

Another aspect concurring to the depiction of the WEF Nexus in Mozambique is represented by the biofuel industry. Biofuels are a rising economic sector in Mozambique, the consequences of which are not fully predictable but still of crucial importance. Indeed, foreign investments in Mozambique increased significantly after the Government created a platform for the evaluation of its biofuel potential. The investments came mostly from Brazilian, Chinese, European and British companies and were directed towards the cultivation of jatropha and sugarcane on the many Mozambican state-owned unused areas (Nielsen et al. 2016). In 2009, Mozambique adopted the National Biofuel Strategy. On one hand, this plan seemed to promote an increase in job opportunities and to foster Mozambican energy self-sufficiency, but it came with several risks concerning food security and land rights. Its strategy indicated the production of second-generation energy crops from marginal land feedstock as a mean to avoid competition with food production. However, because of the complex bureaucracy and the global crisis in 2008, the biofuel production projects started in Mozambique have been few, have underperformed in terms of job creation and, as mentioned earlier, have kept using sugarcane as a feedstock, with well-known consequences on food security (FAO 2016; Nielsen et al. 2016; CARE 2017). Moreover, the increase in the production of highly irrigation-demanding cash crop such as sugar cane, pushed over by foreign investment, is part of a broader framework of large-scale land acquisitions (LSLA) affecting most African countries (Rulli et al. 2018; Chiarelli et al. 2021).

More in general, Mozambican economy has been undergoing a slowdown since the second decade of the twenty-first century, caused mostly by the fall of commodity prices. This, together with the multifaceted picture provided by Mozambique in terms of successes and issues in water governance, agricultural growth, food security and socio-economic development, demonstrates once again the importance of a deeply intersectoral policy framework, able on one hand to be strongly expert-based, and on the other hand to actively and concretely involve all social strata, especially those who have been so far excluded from power positions and decision processes.

References

CARE (2017) Policy analysis. Food security, nutrition, climate change resilience, gender equality and the small-scale farmers—Mozambique. Available via Care International. https://www.care.org.mz/contentimages/policyanalisisMozambiquefinal.pdf. Accessed 8 November 2021

Chiarelli DD, D'Odorico P, Davis KF et al (2021) Large-scale land acquisition as a potential driver of slope instability. L Degrad Dev 32:1773–1785. https://doi.org/10.1002/LDR.3826

Chiarelli DD, Passera C, Rosa L et al (2020) The green and blue crop water requirement WATNEEDS model and its global gridded outputs. Sci Data 7:1–9. https://doi.org/10.1038/s41597-020-00612-0

de Sousa LS, Wambua RM, Raude JM, Mutua BM (2019) Assessment of water flow and sedimentation processes in irrigation schemes for decision-support tool development: a case review for the Chókwè irrigation scheme, Mozambique. AgriEngineering 1:100–118. https://doi.org/10.3390/agriengineering1010008

Donovan C, Tostao E (2010) Staple food prices in Mozambique. Paper presented at the Comesa policy seminar on "Variation in staple food prices: causes, consequence, and policy options", Maputo, 25–26 January 2010

Ducrot R, Leite M, Gentil C et al (2020) Strengthening the capacity of irrigation schemes to cope with flood through improved maintenance: a collaborative approach to analySe the case of Chókwè, Mozambique. Irrig Drain 69:126–138. https://doi.org/10.1002/IRD.2229

FAO (2016) AQUASTAT Country profile—Mozambique. FAO, Rome

FAOSTAT (2018) Food and Agriculture Organization Corporate Statistical Database (FAOSTAT). Available via FAOSTAT. https://www.fao.org/faostat/en/#data. Accessed 8 November 2021

Ferrão J, Bell V, Cardoso LA, Fernandes T (2018) Agriculture and Food Security in Mozambique. J Food, Nutr Agric 7–11. https://doi.org/10.21839/jfna.v1i1.121

Funk C, Peterson P, Landsfeld M et al (2015) The climate hazards infrared precipitation with stations—A new environmental record for monitoring extremes. Sci Data 2:1–21. https://doi.org/10.1038/sdata.2015.66

Lehner B, Verdin K, Jarvis A (2008) New global hydrography derived from spaceborne elevation data. Eos, Transactions, American Geophysical Union. EOS, Trans Am Geophys Union 89:93–104

Ministério da Agricultura (2010) Estratégia de Irrigação. Ministério da Agricultura, Maputo

Ministério da Agricultura (2012) Plano Nacional De Investimento Do Sector Agrário (PNISA) 2013-2017. Ministério da Agricultura, Maputo

Nielsen T, Schhnemann F, McNulty E et al (2016) The food-energy-water security Nexus: definitions, policies, and methods in an application to Malawi and Mozambique. IFPRI Discussion Paper 1480. https://doi.org/10.2139/ssrn.2740663

Pellizzoli R (2010) "Green revolution" for whom? women's access to and use of land in the Mozambique Chckwe irrigation scheme. Rev Afr Polit Econ 37:213–220. https://doi.org/10.1080/03056244.2010.483896

República De Moçambique (1991) Water Law 16/1991. Boletim da República No. 31

República De Moçambique (2007a) Regulamento de Licenças e Concessões de Águas 43/2007a

República De Moçambique (2007b) Estratégia de Segurança Alimentar e Nutricional (ESAN)

República De Moçambique (2016a) Politica de Águas 42/2016a. Boletim da Republica, I Serie, No. 156, 17th Supplement

República De Moçambique (2016b) Programa Nacional de Irrigação (PNI) 43/2016b. Boletim da República, I Série, 17th Supplement, No. 156

República De Moçambique (2010) Multisectoral Action Plan for the Reduction of Chronic Undernutrition in Mozambique (PAMRDC) 2011–2020

República De Moçambique Plano Estratégico para o Desenvolvimento do Sector Agrário (PEDSA) 2011–2020

Rulli MC, Passera C, Chiarelli DD, D'Odorico P (2018) Socio-environmental effects of large-scale land acquisition in Mozambique. Res Dev 377–389. https://doi.org/10.1007/978-3-319-61988-0_29

The World Bank Group (2019) World bank open data. Available via the World Bank. https://data.worldbank.org/. Accedded on 8 November 2021

Veldwisch GJ, Beekman W, Bolding A (2013) Smallholder irrigators, water rights and investments in agriculture: three cases from rural Mozambique. Water Alternatives 6(1):125–141

Water and Climate Change: Water Management in Transboundary River Basins Under Climate Change

Elena Matta and Andrea Castelletti

Abstract Water scarcity is becoming one of the main threats of this century. Population growth, increasing water, energy and food demands, low water use efficiencies and high losses are some of the reasons behind the water crisis worldwide. More than 2.3 billion people are currently living in water-stressed countries and 733 million in highly critical areas. In the Global South, additional pressure is coming from a lack of water management infrastructure, non-adequate sources of drinking water and insufficient sanitation services, unsustainable groundwater withdrawal and unpreparedness to disaster's risks and climate change impacts. In this brief contribution, we first tackle the main challenges of water resource management in the developing countries of the Global South, focusing on sub-Saharan Africa. Beside water scarcity facts, we highlight how these are worsened in transboundary river basins, where cooperation and political agreements are hard to be achieved. The solutions in our hands are sustainable and integrated water management of the multiple uses of water, along with adequate mitigation and adaptation actions to climate change. Afterwards, we describe the expected climate impacts and why African countries are particularly vulnerable to climate change. Finally, an interesting study with a focus on Mozambique is presented, as it shapes well the hard facts in the region, but it also sheds a light on the right direction to take in supporting regional development and cooperative water management.

1 Water Challenges in the Global South

Water is the main driver of this century. Besides being a vital resource for humanity, it is inevitably shrinking while water quality deteriorates, following different spatial and temporal patterns. The global water demand has increased by 600% over the past 100 years, with an annual increment rate of 1.8% (Boretti and Rosa 2019). According to the United Nations (2019), the world population in 2019 accounted for 7.8 billion

E. Matta (✉) · A. Castelletti
Department of Electronics, Information, and Bioengineering, Politecnico di Milano, Milano, Italy
e-mail: elena.matta@polimi.it

© The Author(s), under exclusive license to Springer Nature Switzerland AG 2022
L. Montedoro et al. (eds.), *Territorial Development and Water-Energy-Food Nexus in the Global South*, Research for Development,
https://doi.org/10.1007/978-3-030-96538-9_11

people, while almost 10 billion are predicted in 2050, with more than half of the population growth expected by 2050 to occur in Africa. Two-thirds of the increasing global population (about 4.0 billion people) currently live under conditions of severe water scarcity for at least one month of the year, while half a billion people all year round (Mekonnen and Hoekstra 2016).

Not only population growth—especially in urban areas—but also improving living standards, changing consumption patterns and expansion of irrigated agriculture are challenging water availability worldwide. This water deficit will be particularly severe in Asia and in sub-Saharan Africa, where agriculture is an essential component of the economy and irrigation uses from 75 to 90% of the freshwater derived from rivers and/or pumped from aquifers (Anghileri et al. 2011), as rainfall can be absent for long times of the year.

Climate change is making the situation worse. Wet areas are expected to be wetter, while dry ones are to be drier. Droughts, wildfires and floods are populating the global news and demonstrating how climate change is further challenging the water resources of the planet, increasing water scarcity, especially in developing countries.

Growing population and water consumption, along with climate change impacts, lead not only to water scarcity but also to the deterioration of water quality, and in the worst cases to deaths and diseases by polluted water. Three out of ten people worldwide do not have access to safe drinking water (UN Water 2019) nor to adequate sanitation services (WASH).

It becomes clear how one of the biggest challenges of this century is to meet water demand requirements while protecting and preserving the ecosystems and their water resources. Setting limits to water consumption by river basin, increasing water-use efficiencies and better sharing of the limited freshwater resources are key in reducing the threats posed by water scarcity on biodiversity and human welfare (Mekonnen and Hoekstra 2016).

How can we counteract these dramatic facts? Water resources must be used across national borders and different economic sectors, in a collaborative way and with socio-political agreements among all interested parties. The real win can be achieved with cooperation, not with competition, but by adopting participatory decisions. This is a complex process but the only possible one, which starts with the engagement of multi-sectoral stakeholders and users and continues by accounting for the multiple uses of water and the correspondent uncertainties, e.g. the physical aspects (hydrological, climatological, ecological) of a region. As identified by several experts e.g. (Global Water Partnership 2000; Soncini-Sessa et al. 2007), such actions must be carried out at the river basin level, embracing a holistic process that can be synthesized in the acronym Integrated Water Resources Management (IWRM).

1.1 Integrated Management of WEF Multiple Uses of Water in Transboundary Rivers

Water is a resource, and its development and management are specific to the geographical, historical, cultural and economic contexts of any country. As clearly stated by Jønch-Clausen (2004), the priorities of sustainable integrated water management are highly dependent on the territorial context. One important fact is that a good IWRM process can assist developing countries in achieving the Sustainable Development Goals (SDGs) of the UN Agenda 2030, as it promotes the coordinated development and management of water, land and related resources to maximize the resultant economic and social welfare equitably without compromising the sustainability of vital ecosystems. The three pillars which form the basis of a correct IWRM process are (1) enabling environment of appropriate policies, strategies and legislation for sustainable water resources development and management; (2) putting in place the institutional framework through which policies, strategies and legislation can be implemented; and (3) setting up the management instruments required by these institutions to do their job (as schematized in Fig. 1).

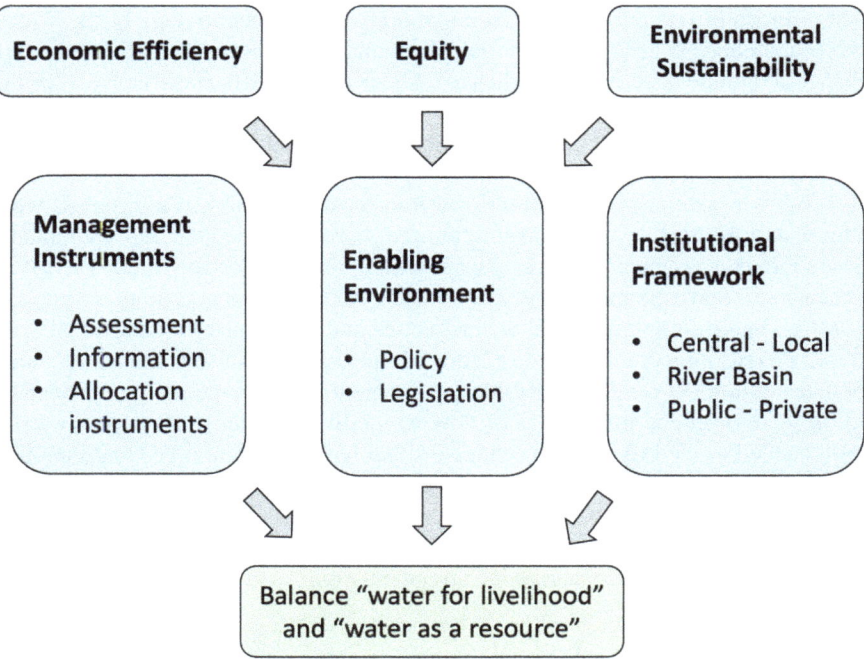

Fig. 1 The three pillars of integrated water resources management: Enabling environment, institutional framework and management instruments, adapted by the authors from the Global Water Partnership (GWP) report of 2004 (Jønch-Clausen 2004)

Starting the engines of such a process is complex, especially in developing countries. It implies a long-term plan and coordination of water, energy, food (WEF) resources and ecosystem preservation, often in transboundary river basins, characterized by scarce finances and policy agreements, besides extreme poverty, indecent sanitation services, gender unbalance and climate change impacts.

In general, water systems are interlinked with food and energy systems and with several sectors, including agriculture, industry and fishery. The availability of fresh water is indeed a limiting factor to food production, energy generation and industrial consumption around the globe and the lack of integration in resource assessments and policymaking leads to inconsistent strategies and inefficient use of resources (Howells et al. 2013; Amaranto et al. 2021). Particularly in sub-Saharan Africa and South-East Asia, this need is supplied by investing in new infrastructure to expand storing and conveying capacity, enhance food security and increase hydropower production. As it is also evident by recent facts around the Grand Ethiopian Renaissance Dam (GERD), the share of water across neighbouring countries, the consequent environmental impacts and social costs can lead to conflicts and critical trade-offs between supply and demand, see e.g. (Wheeler et al. 2020). Even if an agreement on the dam cascade's operation of the Nile River Basin is not yet been met, the attention to cooperative policies and management interventions in the region is improving. Most research efforts are insisting on collaborative operating strategy of the Water Energy and Food (WEF) systems in such transboundary rivers as the only solution to enhance economic benefits and resilience (Wheeler et al. 2016; Basheer et al. 2021). On the other hand, drivers of environmental degradation are growing and measures to preserve water-related ecosystems are generally weak and their scope is not always adequate (NBI 2020).

Changes in natural runoff and inflows from upstream parts of a basin, as well as local and upstream water consumption, are hardly predictable and complicate transboundary agreements. Upstream water availability is indeed an important driver of changes in downstream one but also managing local water consumption plays a big role. Therefore, international water treaties and cooperative management will have an increasingly crucial role to ensure fair management of transboundary water resources (Munia et al. 2020). Beside the problem of hydro-political dependency and the acknowledged importance of flow variability in water treaties, much work is still needed to identify those mechanisms that have been employed and the ones that can succeed. What is promising in Africa—where 64% of the land is covered by transboundary river basins—is that many countries have institutionalized most elements of integrated water resources management to address the multi-sectoral dimension of water security and reduce fragmentation.

2   Climate Change Facts

Climate change represents an additional threat to already stressed water resources and a further challenge to effective long-range policies and integrated management of water resources worldwide, besides it increases uncertainty. This is particularly true for the Global South, especially sub-Saharan Africa, where the targeted key risks are the compounded stress on water resources, reduced crop productivity and livelihood and food security, and vector- and water-borne diseases (IPCC 2014). Disaster risk management, adaptation in technologies and infrastructure, ecosystem-based approaches, basic public health measures and livelihood diversification can reduce vulnerability. Also, there is the need for flexible water supply infrastructure planning, according to different climate and socio-economic projections. Most governments in developing countries are initiating such adaptation measures to reduce the impending climate risks, although efforts to date tend to be isolated and not yet fully effective.

2.1 Climate Impacts

Climate change is posing hard challenges to water resources worldwide, and in particular, developing countries are found to be extremely vulnerable and under-prepared for its impending impacts (Sen Roy 2018). Research centres, universities and experts all over the world have the tools to produce climate projections and provide a plausible representation of our future climate, but they are struggling with a large envelope of uncertainty due to the interconnected modelling cascade and the quality and accessibility of data (Wilby and Dessai 2010). This uncertainty is particularly evident in the Global South, mainly due to the low density of meteorological stations and thus low measurement availability (Diallo et al. 2015). In those regions, it is hard to find a clear and consistent pattern, especially looking at precipitation projections, which are showing high temporal and spatial variability (Hasan et al. 2019; Ogega et al. 2020).

Some of the most evident climate impacts include decreasing trends in precipitation accompanied by increasing trends in temperatures and extreme weather events. For example, in sub-Saharan Africa, regional climate models can deliver reliable simulations of rainfall, but with noteworthy biases over different subregions (Nikulin et al. 2018). Climate change is projected to increase the interannual variability of rainfall, which can cause devastating droughts and floods. In particular, the eastern parts of Africa are found to be highly prone to changes in climate and climate extremes, and more frequent droughts, floods and heavy rainstorms are projected in the future (Gebrechorkos et al. 2019). Some studies highlighted that the variability in rainfall can be linked to large-scale climate variability, including the El Niño Southern Oscillation (ENSO) (Endris et al. 2013; Giuliani et al. 2019). Changes in sea surface temperature can also affect the rainfall amount (e.g. decrease during the rainy season) by changing wind patterns and moisture fluxes (Endris et al. 2019).

The Intergovernmental Panel on Climate Change (IPCC) sixth assessment report—not yet completed (IPCC 2021) states that the likely range of human-induced change in global surface temperature in 2010–2019 relative to 1850–1900 is 0.8 °C–1.3 °C, with a central estimate of 1.07 °C, while the likely range of the change attributable to natural forcing is only –0.1 °C to +0.1 °C. The bad news is that the expected warming is worsened in Africa since regional climate response to these targets is much faster there than the global warming average. Of particular concern in Africa are heatwaves and their duration, as well as rainfall extremes because of the devastating impact they have across natural and socio-economic systems (Lennard et al. 2018). For instance, it is observed that the most severe effects of global warming will be related to the frequency and severity of extreme events (Dosio 2017). As reported in the report by World Meteorological Organization (WMO 2019), the year 2019 was among the three warmest years on record for the continent. That trend is expected to continue, along with the increase of sea level rise.

Consequently to the temperature extremes, the drought risk is larger and it shows—on the one hand—that already dry and arid regions as the northern African countries might experience an aggravating drought hazard, but—on the other hand— the drought risk ratio is found to be the highest in central African countries because of vulnerability and population rise in that region (Ahmadalipour et al. 2019). Climate change adaptation is thus dramatically urged. One effective mitigation action is to keep population growth under control, to improve socio-economic vulnerability and reduce potential exposure to drought (Ahmadalipour et al. 2019; Parkes et al. 2019).

Beside assessing climate forthcoming effects on hydrology and socio-economy, a large part of the research focuses on climate change impacts on dams operations and water resources management along transboundary river basins, which can support resolving the conflicts around the water rights and availability, as well as promoting cooperative management as winning action by all sharing countries (Jeuland and Whittington 2014; Liersch et al. 2017; Parkes et al. 2019).

It is therefore evident how climate change is exacerbating the water issue by posing serious threats for human health, food and water security and socio-economic development in developing countries, as schematized in Fig. 2. In this context, science-based climate information is the foundation to build climate-resilient policies, adaptation and mitigation actions.

2.2 Why is Africa Extremely Vulnerable to Climate Change?

According to official international reports (IPCC 2014; World Meteorological Organization (WMO) 2019), Africa is found to be extremely vulnerable to climate change for several reasons, summarized in the following:

- Agriculture is the backbone of Africa's economy.
- It has a weak adaptive capacity and high dependence on ecosystem goods for livelihoods, and less developed agricultural production systems (Sub-Saharan Africa accounts for 95% of rain-fed agriculture globally).
- It is the hotspot for climate variability and change: seven out of ten countries have been declared as most vulnerable to climate change are in Africa, with Mozambique in the first place.

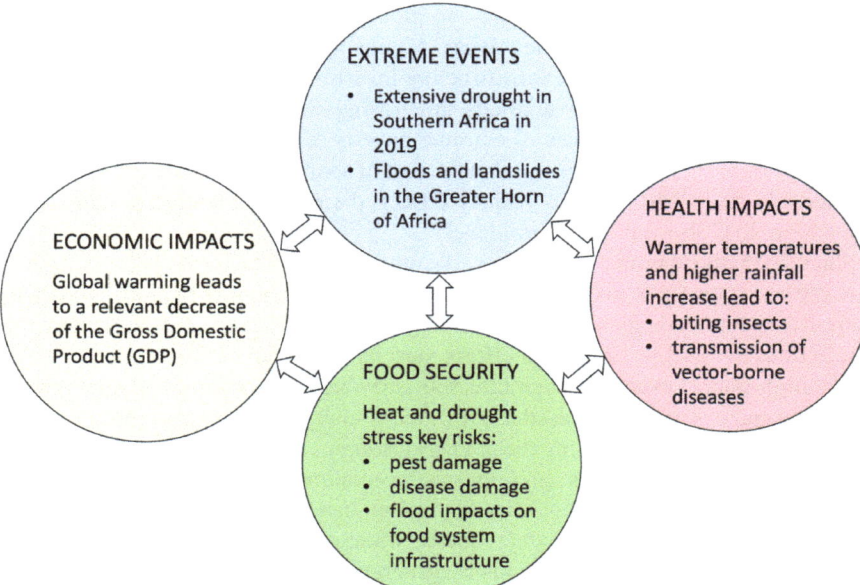

Fig. 2 Climate change impacts and challenges considering different sectors, inspired by the concepts included in the World Meteorological Organization (WMO) climate report (World Meteorological Organization (WMO) 2019)

- Warming scenarios in Africa risk having devastating effects on crop production and food security. Also, a large share of agriculture in GDP and employment adds to vulnerability, as do other weather-sensitive activities, such as herding and fishing, leading to income losses and increased food insecurity.
- It is one of the regions facing the largest capacity gaps regarding climate services, with the least developed land-based observation network of all continents.
- It experiences a compounded stress on water resources, such as population growth, international cooperation of transboundary rivers, new water infrastructure management, ecosystem services and climate change.
- It is prone to vector- and water-borne diseases.
- The gender rights are unbalanced, as many women are working in the field and have though no access to climate services, which means that they are less prepared for risks.
- It lacks risk management and prevention, such as multi-hazards early warning systems.
- African commitments to counteract climate change and its impacts are depending on receiving adequate financial, technical and capacity-building support.
- There are 63 transboundary river basins in Africa, crossing countries with different climatic, hydrological, topological and economical characteristics.

Despite having contributed the least to global warming and having the lowest greenhouse gas emissions, Africa faces exponential collateral damage, posing systemic risks to its economies, infrastructure investments, water and food systems, public health, agriculture and livelihoods, threatening to undo its modest development gains and slip into higher levels of extreme poverty. The United Nations Environment Programme (UNEP) estimates that the cost of adapting to climate change across Africa could reach $50 billion a year by 2050 if the global temperature increase is kept within 2 °C above preindustrial levels.

Since 2015, the Nationally Determined Contributions (NDCs) to the Paris Agreement represent the main instrument for guiding policy responses to climate change, where Africa has made great efforts in driving the global climate agenda. Among the main actions in this direction, great efforts must be surely pursued to build resilience against high-impact events through effective early warning systems and appropriate prevention and risk management strategies. Additionally, enhancing promotive socioeconomic growth, implementing clean energy sources (wind, solar) and embracing innovative farming techniques can improve the economy and reduce emissions. For instance, solar-powered efficient micro-irrigation can increase farm-level incomes, improving yields by up to 300% and reducing water usage by up to 90%, while at the same time offsetting carbon emissions by generating up to 250 kW of clean energy. Another essential point to consider is that women still represent a large percentage of the world poor in developing countries. They are also active in agriculture, and they often have no access to climate services. Promoting gender balance and ensuring equal rights to men and women is fundamental to enhance the individual resilience and adaptive capacity of all individuals, who can finally be provided with access to climate services and risk warning systems.

3 Zoom in Mozambique

Mozambique is one of the poorest countries in the world, with half the urban population living below the national poverty line and only a quarter have access to piped water. The growth of population and urban migration is worsening the condition of water and sanitation services in the country. Despite significant progress over the years, only half of Mozambicans have access to improved water supply and less than a quarter (one in five) use improved sanitation facilities (UNICEF 2021). Irrigated agriculture is important for poverty alleviation and improving food security in Sub-Saharan Africa but has often failed. At the same time, strong population growth continues to increase the demand for water for urban and other non-agricultural users (Pittock and Blessington 2021).

Beside the fear of being unable to supply the increasing water demands, Mozambique is likely to experience severe water quality issues in addition to availability problems. There are studies in the region that proved how operation rules of reservoirs, involving water intakes at different levels, could mitigate the consequences for downstream water quality (Calamita et al. 2021). Finding strategies to mitigate anthropogenic stress to the ecosystem is of high importance, especially for ecosystems where the climate is already imposing a high impact. The consequences of hydrological and water quality alterations, river damming at low latitudes can affect the food availability for local populations. Another aspect to consider is how to design and maintain environmental flows in the Mozambiquan transboundary rivers, not only as a fundamental element for ecosystem sustainability but also an additional challenge to water resource management (Nhassengo et al. 2021).

3.1 A Multi-objective WEF Study in the Umbeluzi River Basin, Mozambique

An interesting study was conducted in the Lower Umbeluzi River Basin, in Mozambique, as it represents an archetypal example of a complex water system, where water scarcity combined with climatic variability, uncontrolled urbanization and agricultural expansion is substantially stressing the agricultural, energy and urban sectors (Amaranto et al. 2021). The Lower Umbeluzi River Basin is a highly regulated, fast-evolving sub-Saharan hydrosystem. The river flows in the Pequenos Libombos reservoir, about 45 km upstream before reaching its delta in Maputo Bay. The operation of the Pequenos Libombos Dam is aimed to balance hydropower production, urban uses and irrigation in a very challenging context. Climate impacts such as long-lasting droughts and tropical cyclones in the region are aggravating conflicts among the urban, agricultural and energy sectors. For these reasons, the World Bank funded the Greater Maputo Water Supply Expansion Project, a sequence of infrastructural interventions started in 2013, with the scope to supply the city of Maputo with an additional inflow from the Sabie River Basin.

The uncertain evolution of climatic, agricultural, infrastructural and social patterns in the area calls for assisting policymakers in developing robust reservoir operating policies, to better assess the main sources of vulnerability of sustainable water supply and to quantify such uncertainty sources across the different sectors. In the research we are reporting here, such an analysis was enabled by the development of an integrated decision-analytic framework that combined optimization, robustness, sensitivity and uncertainty analysis, based on multi-objective optimization algorithms. Figure 3 is an effective example of different compromise solutions of a typical multi-stakeholder dynamics. Looking at different objectives and stakeholders (categorized as upstream irrigation, hydropower, urban deficit and irrigation downstream), the most conflicting ones under the explored baseline scenarios resulted to be hydropower and upstream irrigation. The irrigation demand cannot directly benefit from hydropower generation (which is instead positive for the energy sector), but it withdraws water from the reservoir to supply the crop requirements (though positive for the agricultural sector). Simulating different agricultural water demand scenarios, the irrigation downstream appeared to be the least vulnerable stakeholder to deep uncertain scenarios (while hydropower is the most vulnerable). The robust policy

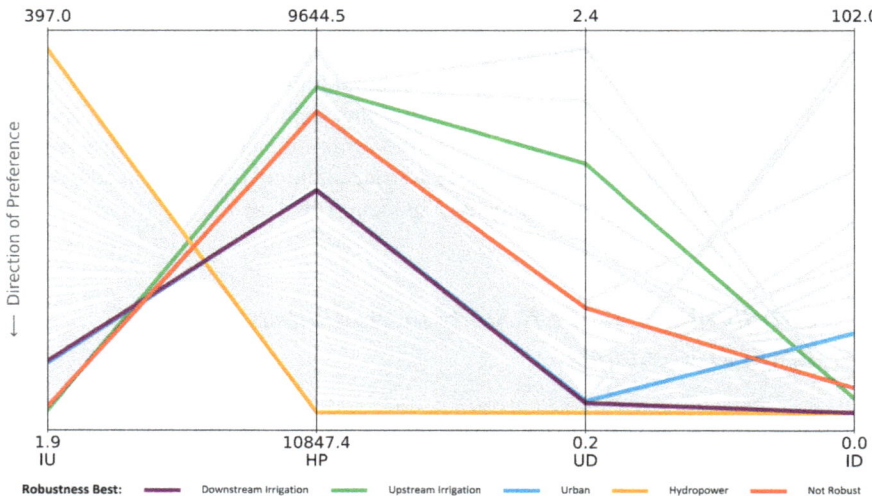

Fig. 3 Policy performance across upstream irrigation (IU), hydropower (HP), urban deficit (UD) and irrigation downstream (ID), shown on the horizontal axis of the graph. The direction of preference is shown on the vertical axis, increasing as we move from the top to the bottom. The coloured lines indicate the most robust policies, except for the red one, which represents a non-robust one (all others are in light grey) (Amaranto et al. 2021).

for downstream irrigation ensures that agricultural production can be sustained with a near-zero deficit for most of the cases explored (as shown by the purple line in Fig. 3).

The outcomes of this research provide important insights into the robustness and vulnerability of reservoir operation to exogenous perturbations in managing multiple, conflicting objectives. It was concluded that robust policies turn to be vulnerable only to hydrological perturbations and can sustain most of the population growth and agricultural expansion scenarios, while infrastructural interventions become crucial only in extreme drought conditions.

4 Conclusive Remarks

This brief contribution intends to highlight the complex interconnected nature of water-related issues in the Global South, which often results in competing water demand and water resources, as well as water pollution. It is thus extremely important to concentrate international efforts to work on the interactions between climate change, population and economic growth, to perform an equitable and sustainable use and management of water resources for poverty alleviation, socio-economic development, regional cooperation and environment preservation. Future studies should also focus on developing better procedures to estimate environmental flow requirements per catchment, including the effects of climate change in their assessments, to protect the delicate freshwater ecosystem.

Water is a finite and vulnerable resource, and it urges not only continuous scientific research advances, but also resilient policies, development plans and financial support, especially in the countries of the Global South, where the hope lies only in a concrete change.

References

Ahmadalipour A, Moradkhani H, Castelletti A, Magliocca N (2019) Future drought risk in Africa: integrating vulnerability, climate change, and population growth. Sci Total Environ 662:672–686. https://doi.org/10.1016/j.scitotenv.2019.01.278

Amaranto A, Juizo D, Castelletti A (2021) Disentangling sources of future uncertainties for water management in sub-saharan river basins. Hydrol Earth Syst Sci. https://doi.org/10.5194/hess-202 1-40

Anghileri D, Pianosi F, Soncini-Sessa R (2011) A framework for the quantitative assessment of climate change impacts on water-related activities at the basin scale. Hydrol Earth Syst Sci 15:2025–2038. https://doi.org/10.5194/hess-15-2025-2011

Basheer M, Nechifor V, Calzadilla A, Siddig K, Etichia M, Whittington D, Hulme D, Harou JJ (2021) Collaborative management of the grand Ethiopian renaissance dam increases economic benefits and resilience. Nat Commun 2021 121 12(1):1–12. https://doi.org/10.1038/s41467-021-25877-w

Boretti A, Rosa L (2019) Reassessing the projections of the world water development report. npj Clean Water 2019 21 2(1):1–6. https://doi.org/10.1038/s41545-019-0039-9

Calamita E, Vanzo D, Wehrli B, Schmid M (2021) Lake modelling reveals management opportunities for improving water quality downstream of transboundary tropical dams. Water Resour Res :e2020WR027465. https://doi.org/10.1029/2020WR027465

Diallo I, Giorgi F, Sukumaran S, Stordal F, Giuliani G (2015) Evaluation of RegCM4 driven by CAM4 over Southern Africa: mean climatology, interannual variability and daily extremes of wet season temperature and precipitation. Theor Appl Climatol 121(3–4):749–766. https://doi.org/10.1007/s00704-014-1260-6

Dosio A (2017) Projection of temperature and heat waves for Africa with an ensemble of CORDEX regional climate models. Clim Dyn 49(1–2):493–519. https://doi.org/10.1007/s00382-016-3355-5

Endris HS, Lennard C, Hewitson B, Dosio A, Nikulin G, Artan GA (2019) Future changes in rainfall associated with ENSO, IOD and changes in the mean state over Eastern Africa. Clim Dyn 52(3–4):2029–2053. https://doi.org/10.1007/s00382-018-4239-7

Endris HS, Omondi P, Jain S, Lennard C, Hewitson B, Chang'a L, Awange JL, Dosio A, Ketiem P, Nikulin G, Panitz HJ, Büchner M, Stordal F, Tazalika L, (2013) Assessment of the performance of CORDEX regional climate models in simulating East African rainfall. J Clim 26(21):8453–8475. https://doi.org/10.1175/JCLI-D-12-00708.1

Gebrechorkos SH, Hülsmann S, Bernhofer C (2019) Long-term trends in rainfall and temperature using high-resolution climate datasets in East Africa. Sci Rep 9(1):1–9. https://doi.org/10.1038/s41598-019-47933-8

Giuliani M, Zaniolo M, Castelletti A, Davoli G, Block P (2019) Detecting the state of the climate system via artificial intelligence to improve seasonal forecasts and inform reservoir operations. Water Resour Res 55(11):9133–9147. https://doi.org/10.1029/2019WR025035

Global Water Partnership (2000) Integrated water resources management global water partnership technical advisory committee (TAC). Stockholm, Sweden

Hasan E, Tarhule A, Zume JT, Kirstetter PE (2019) +50 years of terrestrial hydroclimatic variability in Africa's transboundary waters. Sci Rep 9(1):1–12. https://doi.org/10.1038/s41598-019-488 13-x

Howells M, Hermann S, Welsch M, Bazilian M, Segerström R, Alfstad T, Gielen D, Rogner H, Fischer G, van Velthuizen H, Wiberg D, Young C, Roehrl RA, Mueller A, Steduto P, Ramma I (2013) Integrated analysis of climate change, land-use, energy and water strategies. Nat Clim Chang 2013 37 3(7):621–626. https://doi.org/10.1038/nclimate1789

IPCC (2014) Climate change 2014: synthesis report. Contribution of Working Groups I, II and III to the Fifth Assessment Report of the Intergovernmental Panel on Climate Change. Gian-Kasper Plattner, Geneva, Switzerland

IPCC (2021) Sixth assessment Report. https://www.ipcc.ch/report/ar6/wg1/#TS. Accessed 10 Oct 2021

Jeuland M, Whittington D (2014) Water resources planning under climate change: assessing the robustness of real options for the Blue Nile. Water Resour Res 50(3):2086–2107. https://doi.org/10.1002/2013WR013705

Jønch-Clausen T (2004) Integrated Water Resources Management (IWRM) and Water Efficiency Plans by 2005. Why, What and How?

Lennard CJ, Nikulin G, Dosio A, Moufouma-Okia W (2018) On the need for regional climate information over Africa under varying levels of global warming. Environ Res Lett 13(6). https://doi.org/10.1088/1748-9326/aab2b4

Liersch S, Koch H, Hattermann FF (2017) Management scenarios of the grand Ethiopian renaissance dam and their impacts under recent and future climates. Water 9(10):728. https://doi.org/10.3390/w9100728

Mekonnen MM, Hoekstra AY (2016) Sustainability: four billion people facing severe water scarcity. Sci Adv 2(2). https://doi.org/10.1126/SCIADV.1500323

Munia HA, Guillaume JHA, Wada Y, Veldkamp T, Virkki V, Kummu M (2020) Future transboundary water stress and its drivers under climate change: a global study. Earth's Futur 8(7). https://doi.org/10.1029/2019EF001321

NBI - Nile Basin Initiative (2020) State of the Nile river basin. NBI, Entebbe

Nhassengo OSZ, Somura H, Wolfe J (2021) Environmental flow sustainability in the lower Limpopo river basin. Mozambique. J Hydrol Reg Stud 36. https://doi.org/10.1016/J.EJRH.2021.100843

Nikulin G, Lennard C, Dosio A, Kjellström E, Chen Y, Hansler A, Kupiainen M, Laprise R, Mariotti L, Maule CF, Van Meijgaard E, Panitz HJ, Scinocca JF, Somot S (2018) The effects of 1.5 and 2 degrees of global warming on Africa in the CORDEX ensemble. Environ Res Lett 13(6):065003. https://doi.org/10.1088/1748-9326/aab1b1

Ogega OM, Koske J, Kung'u JB, Scoccimarro E, Endris HS, Mistry MN, (2020) Heavy precipitation events over East Africa in a changing climate: results from CORDEX RCMs. Clim Dyn 55(3–4):993–1009. https://doi.org/10.1007/s00382-020-05309-z

Parkes B, Cronin J, Dessens O, Sultan B (2019) Climate change in Africa: costs of mitigating heat stress. Clim Change 154(3–4):461–476. https://doi.org/10.1007/s10584-019-02405-w

Pittock J, Blessington L (2021) Water markets in Africa: an analysis of Mozambique, Tanzania and Zimbabwe. Water Mark :50–63. doi:https://doi.org/10.4337/9781788976930.00012

Sen Roy S (2018) Climate change in the global south: trends and spatial patterns. Springer Climate. Springer, Cham

Soncini-Sessa R, Weber E, Castelletti A (2007) Integrated and participatory water resources management-theory. Elsevier Science Pub. Co

UN Water (2019) The United Nations World Water Development Report. Leaving no one behind. Perugia, Italy

UNICEF Water, sanitation and hygiene (WASH) | UNICEF Mozambique (2021) https://www.unicef.org/mozambique/en/water-sanitation-and-hygiene-wash. Accessed 22 Sep 2021

United Nations (2019) World population prospects 2019, volume II: Demographic Profiles

Wheeler KG, Basheer M, Mekonnen ZT, Eltoum SO, Mersha A, Abdo GM, Zagona EA, Hall JW, Dadson SJ (2016) Cooperative filling approaches for the Grand Ethiopian Renaissance Dam. Water Int 41(4):611–634. https://doi.org/10.1080/02508060.2016.1177698

Wheeler KG, Jeuland M, Hall JW, Zagona E, Whittington D (2020) Understanding and managing new risks on the Nile with the grand Ethiopian renaissance dam. Nat Commun 11(1):1–9. https://doi.org/10.1038/s41467-020-19089-x

Wilby RL, Dessai S (2010) Robust adaptation to climate change. Weather 65(7):180–185. https://doi.org/10.1002/wea.543

World Meteorological Organization (WMO) (2019) State of the climate in Africa. Switzerland, Geneva

Energy: an Essential Asset for the Development of the African Continent

Matteo Vincenzo Rocco and Lorenzo Rinaldi

Abstract Africa is possibly the richest continent in terms of primary resources, from fossil fuels to materials and solar radiation. However, on the other hand, despite the progressive economic improvements, it is still the poorest region at the global level, and more than half of its population is still employed in the primary sector. To reach better living conditions, access to electricity and to modern clean cooking solutions play a crucial role. This section provides a synthetic picture regarding the relevance of the nexus between energy access and socio-economic development. In conclusions, future pathways on how energy conditions in African countries are expected to evolve according to the main authoritative scenarios are discussed.

1 Energy as a Basic Need: A Focus on Africa

The African continent is a land full and rich or primary energy resources. Among African countries there are some of the world largest oil producers, such as Nigeria and Angola, while Africa itself is the second net exporter of oil after Middle East, and oil proven reserves are noticeable, around one fifth of the Middle East ones. It is also the second net exporter of gas: Mozambique and Tanzania could possibly exploit recently discovered offshore reserves to place themselves in a strategic position in the natural gas international trades. The development of new coal resources is hindered by their remoteness and the lack of suitable railway and port infrastructure. Also due to its low cost, coal remains an asset in African societies, therefore the largest portion of coal production is consumed domestically, with Maputo (the capital of Mozambique) being the only major export node towards foreign markets. Renewable energy potential is noticeable throughout the whole continent, with the highest solar energy availability in the world, but also considerable wind and hydro potentials.

However, still in 2020, sub-Saharan Africa is the poorest region at the global level, its economy is not growing at the same pace as other developing regions and more

M. V. Rocco (✉) · L. Rinaldi
Department of Energy, Politecnico Di Milano, Milan, Italy
e-mail: matteovincenzo.rocco@polimi.it

than half of the population is employed in agriculture, which often does not produce economic value since it is performed for self-subsistence (IEA 2019). Regarding access to energy, there are debates about the proper definition of *modern energy access*, but some elements are conventionally accepted, namely access to electricity and access to clean cooking solutions. Around 580 million people (roughly half of its total population) lacks access to electricity (IEA 2020), mainly due to insufficient infrastructure and unreliability of the provision service. Sub-Saharan Africa is, in fact, faces the highest challenges in terms of electricity grid extension, as well as for telecommunication and transportation infrastructures. Despite access to electricity rate is promisingly growing, access to clean cooking solutions still represent a hard obstacle towards the achievement of sustainable development goals. Around 900 million people rely on traditional fuels and cooking stoves, mainly fired by fuelwood or charcoal. In urban areas, also Liquid Propane Gas (LPG) and electric stoves are present in a marginal proportion, which becomes even smaller when considering that the greatest share of the population lives in rural areas. Accounting to IEA (International Energy Agency), the demand of bioenergy will rise by 40% in the next two decades, which may lead to environmental stress in terms of reduction of forest land, as well as negative health impact related to indoor air pollution. The shift from traditional biomass to modern fuels at domestic level, however, is not only linked to income rise, since cultural and social aspect of food is often neglected. It has been noticed, in fact, that the principle of the energy ladder is not always true when dealing with cooking habits: while households tend to increase their electricity consumption according to their income level, the do not switch fully to modern fuels for cooking but they tend to rely on different fuel sources for the preparation of different meals (Nansaior et al. 2011; Van der Kroon et al. 2013).

These general statements explain the considerably lower-than-average energy demand per capita, which is roughly one third of the global average, as well as to high energy and intensity, expressed in terms of unit of energy produced per unit of GDP. In fact, households can typically spend 20–25% of their income on kerosene for lighting service, which can be 150 times higher than the cost to be sustained to have the same need fulfilled by means of incandescent bulbs and 600 times higher with respect to LED bulbs. Regarding the environmental dimension, CO_2 emissions per unit GDP intensity is also well above the global average; however, sub-Saharan Africa represents only a small contribution to global energy-related CO_2 emissions, accounting for merely 3% of the total in 2040, but is on the front line when it comes to the potential impacts of a changing climate. In particular, hydropower prospects can be affected by changing patterns of rainfall and run-off. The fuelwood and charcoal sectors operate largely outside the formal economy, meaning that policymakers have few levers to promote more sustainable forestry.

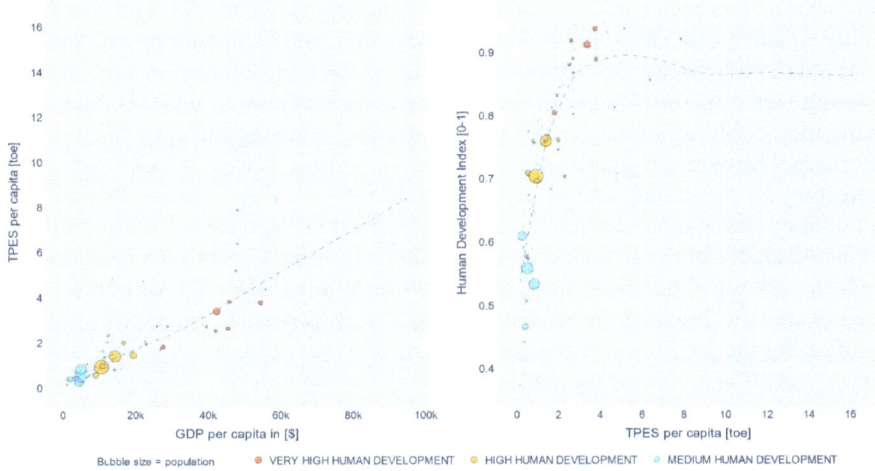

Fig. 1 Graphical correlation among energy, economic and social indicators. *Source* Authors elaboration from UN, World Bank and IEA data (UN 2019; World Bank 2021, IEA 2021)

2 The Nexus Between Energy and Development

The crucial role of energy for development has been strongly investigated within the scientific literature. As anticipated in the previous paragraph, it is important to highlight the cross-cutting relevance of energy along different dimensions, moving from the economic sphere to the social and geopolitical ones.

Figure 1 displays the correlations between selected energy and socio-economic indicators. While each bubble represents a different country, classified by very high, high and medium human development, the chart on the left shows the strong relationship between the produced energy and the GDP per capita. On the right side, it is also clear how the level of energy produced per capita affects the human development index, which is a proxy of different development dimensions, including education and life expectancy.

This synthetic yet comprehensive representation emphasizes the key role of energy for development, therefore Africa has been called to design proper energy policies which will be responsible for designing sustainable pathways to achieve energy needs but also environmental, economic, and social targets. Since the release of Agenda 2030 (UN 2015), one of the Sustainable Development Goals (SDG) was fully dedicated to achieve universal affordable and clean energy access at global level, and starting from the Global Conference on Rural Energy Access[1] an important focus has been given to the nexus between energy and other sustainable development factors, where it has been stated that investments in the energy sectors can create synergies

---

[1] Which took place in Addis Ababa on the 4–6 December 2013. For more details please see: https://sustainabledevelopment.un.org/index.php?page=view&nr=489&type=13&menu=1634. Accessed on 14 December 2021.

with other sectors such as health, food security, poverty eradication, wealth creation, gender equality, education, environmental protection, and humanitarian assistance.

Access to electricity can create opportunities for both women and the youth, fostering new businesses within communities and/or increasing their productivity, giving the possibility to the youth to study at night, allowing the electrification of schools and hospitals, triggering the market development as well as making it more interesting for the private sector investments. Moreover, it is estimated that indoor air pollution due to combustion of traditional biomass is responsible for four million premature deaths per year, a number that could be lowered significantly with access to clean cooking. Therefore, it is clear how the role of SDG 7 (Affordable and clean energy) is crucial to reach the other SDGs, in particular energy can impact significantly the path towards the achievement of SDG 1 (No poverty) and SDG 8 (decent work and economic growth) since energy represents a way out of poverty and an opportunity for job creation and income generation. Access to clean energy, as already mentioned, would improve bad health conditions related to indoor air quality, facilitating the path for SDG 3 (Good health and well-being). In general, many other SDGs may be positively influenced by acting on energy-related leverages.

3   Future Expectations

Drawing from such considerations, it is important to understand how sub-Saharan Africa will play its role within the energy market, in consideration of the growing share of energy and mineral resources which provides to the global system. And how this will contribute to overcome localized challenges in terms of energy access.

According to the IEA scenarios, sub-Saharan Africa energy system is expected to expand rapidly to 2040, but so do the demands placed upon it. The power mix becomes more diverse, with coal (mainly South Africa) and hydropower (all regions) being joined by greater use of gas (Nigeria, Mozambique, Tanzania), solar (notably in South Africa and Nigeria) and geothermal (East Africa). Within the same timeframe, the share of renewables in total capacity more than doubles to arrive at 40%. Total power sector investment averages around $46 billion per year, with just over half of it in transmission and distribution. The economy quadruples in size, the population nearly doubles (to 1.75 billion) and energy demand grows by around 80%. The capacity and efficiency of the system improves, and access to modern energy services grows; however, many of the existing energy challenges are only partly overcome. Bioenergy demand grows by 40% in absolute terms by 2040, exacerbating stress on the forestry stock. However, the share of bioenergy in the energy mix declines from above 60% to below 50%. Oil demand more than doubles and becomes the second largest fuel in the mix, overtaking coal. Natural gas use grows by nearly 6% per year and energy exports are drawn increasingly towards Asian markets. Crude oil net exports decline partly due to a greater share being refined and consumed domestically. Rising gas output from Mozambique and Tanzania increases sub-Saharan Liquified Natural Gas (LNG) export, and Mozambique also joins South Africa as a key coal exporter.

However, in 2040, still 600–800 billion people are expected to lack of clean energy access.

4 Closing Remarks

The two faces of energy, its positive and negative aspects, are more clearly visible in sub-Saharan Africa than in any other part of the world: energy is a critically important enabler of social and economic development and a source of revenue for much-needed investment in infrastructure and other purposes, however where electricity is lacking or resources are poorly managed, energy can also become a source of division, conflict, environmental degradation, poverty. African energy system expands at a good pace, but still struggles to meet the demand. And while access to modern energy services grows, hundreds of millions, particularly in rural communities, are left without. To move towards a better-functioning energy sector, it is requested a step change in investments towards domestic energy supply: increasing investment in the sub-Saharan power sector is essential to bring this equation into line with the regional energy needs. Also, it is necessary to better manage the domestic primary resources: sub-Saharan Africa is rich in energy resources, both fossil fuel and renewable, but the opportunities that these offer to support economic growth are often missed due to the lack of essential infrastructure and robust political frameworks. In the end, expanding cross-border trade can be a very cost-effective way to increase the reliability and affordability of energy supply, but this is often hindered in practice by a range of technical and political barriers.

References

IEA—International Energy Agency (2019) Africa energy outlook. Available via IEA. https://iea.blob.core.windows.net/assets/2f7b6170-d616-4dd7-a7ca-a65a3a332fc1/Africa_Energy_Outlook_2019.pdf, Accessed 22 September 2021

IEA—International Energy Agency (2020) World energy outlook. Available via IEA. https://iea.blob.core.windows.net/assets/888004cf-1a38-4716-9e0c-3b0e3fdbf609/WorldEnergyOutlook2021.pdf, Accessed 22 September 2021

IEA—International Energy Agency (2021) IEA data and statistics. Available via IEA. https://www.iea.org/data-and-statistics, Accessed 22 Sept 2021

Nansaior A, Patanothai A, Rambo AT, Simaraks S (2011) Climbing the energy ladder or diversifying energy sources? The continuing importance of household use of biomass energy in urbanizing communities in Northeast Thailand. Biomass Bioenerg 35(10):4180–4188. https://doi.org/10.1016/j.biombioe.2011.06.046

UN—United Nations (2015) Transforming our world: the 2030 Agenda for Sustainable Development. United Nations, New York

UN—United Nations, Department of Economic and Social Affairs (2019) World population prospects 2019. Available by UN. https://population.un.org/wpp/Publications/Files/WPP2019_Highlights.pdf, Accessed 22 Sept 2021

Van der Kroon B, Brouwer R, van Beukering PJH (2013) The energy ladder: Theoretical myth or empirical truth? Results from a meta-analysis. Renew Sustain Energy Rev 20:504–513. https://doi.org/10.1016/j.rser.2012.11.045

WB—World Bank (2021) Wolrd bank open data. Available via Word Bank. https://data.worldbank.org/, Accessed 22 Sept 2021

Environment: A Bioclimatic Approach to Urban and Architectural Design in Sub-Saharan African Cities

Valentina Dessì

Abstract Drawn on shared understandings of what "environment" might mean across different disciplines and making reference to the most recent literature on bioclimatic and sustainable urban and architectural design, this chapter aims at providing some general guidelines on how to work with local climate conditions and materials to achieve more sustainable and climate-sensitive building design. This is particularly crucial when dealing with fast-changing urban areas—especially in the sub-Saharan African context—where the climate crisis combined with unprecedented urbanization trends are increasing the pressure on natural environments and resources becoming more and more endangered and scarce. These challenges need to be addressed through a conscious and a cross-scalar approach towards the built environment. The case of Mozambique is adopted as a testbed for the combined use of traditional building techniques and innovative bioclimatic methods adopting basic energy flows—solar irradiation, winds and air humidity, among many others—as main guiding parameters to achieve a better comfort both in new or renovated buildings.

1 Introduction

The definitions provided by different sources of the term "Environment" share the consideration that this concept is a synthesis of several factors that determine the characteristics of a place.

The Cambridge dictionary, defining that the environment is represented by the set of 'air, water, and land in or on which people, animals, and plants live'[1] does not specify, unlike the Merriam Webster dictionary, that the environment modifies the

---

[1] https://dictionary.cambridge.org/it/dizionario/inglese/environment (Accessed 10 December 2021).

V. Dessì (✉)
Department of Architecture and Urban Studies, Politecnico di Milano, Milan, Italy
e-mail: valentina.dessi@polimi.it

way an individual or an ecological community lives. The most relevant to this context definition of Merriam Webster dictionary defines the environment as 'the complex of physical, chemical, and biotic factors (such as climate, soil, and living things) that act upon an organism or an ecological community and ultimately determine its form and survival'.[2] Air–water–soil, or climate–soil are the interface elements we deal with: the environment influences the possibility of organisms, animals and plants to exist and organize themselves to survive.

Our survival in a habitat is allowed by the environment. At the same time, it is also true that we have been able to modify the environment by controlling its elements to enable our adaptation to it. Unfortunately, our ability to change the environment has in many cases led to its deterioration: excessive consumption of soil, water contamination, air pollution resulting in climate change and impacts that change our relationship with the environment itself (IPCC 2014).

The concept of "anthropocene" (Crutzen and Stoermer 2000) reminds us that human beings are becoming the major ecological force. For this reason, we are required to take responsibility for the environment.

Besides the policies necessary for protecting the environment and focusing on the topic from the urban and architectural designer's point of view, it is appropriate to include in the definition the possibility to act on the environment itself, with proper methods and considering the factors that configure it. It is a design approach that has to be able to respect and enhance the environment and make it hospitable, liveable and comfortable to the people and living organisms that reside in it and represent the ecological community.

Different approaches are related to the transformation of the environment, such as bioclimatic architecture, sustainable design, design with nature, climate-responsive, etc. They are not mutually exclusive and often pursue the same goal of guaranteeing acceptable comfort conditions for people and maintaining balance with the material and intangible resources in the context. In some cases, the resources can be implemented to improve microclimatic conditions both indoor and outdoor, but sometimes they need to be preserved and protected. Bioclimatic, as defined by the Collins dictionary (2021), is an adjective 'concerning the relationship between climate and living organism'[3] (including buildings). The bioclimatic design approach aims to maximize comfort level by optimizing the sustainability and enhancing the passive solutions, connected to local climate, due to the design of the building envelopes that helps to improve (in the cold season) and control and limit (in the warm-hot season) the heat.

The environmental approach to design—which crosses the different scales of the city and feeds on its interactions—is a cross-disciplinary path that includes the knowledge of the context and the elements, such as materials, vegetation, water, as well as energy flows (i.e. solar radiation), winds, etc. It is also the search for a balance between the elements that find in the urban and architectural project their synthesis.

---

[2] https://www.merriam-webster.com/dictionary/environment (Accessed 10 December 2021).

[3] https://www.collinsdictionary.com/it/dizionario/inglese/bioclimatic (Accessed 10 December 2021).

2 African Environment and Mozambique

In the broad context of living in Mozambique and more generally in sub-Saharan Africa, the environment is a concept that includes aspects of fragility, especially concerning a series of epochal changes taking place, including demographic growth of the population and the rural-to-urban migrations.

In 1990, just over 30% of the African population used to live in cities; by 2035, the percentage could rise to 50%, according to the United Nations (UN 2017). This is followed by an immediate and direct consequence of the informal expansion of the small-, medium-sized towns and megacities. The resulting population growth—in general, and more specifically in urbanized areas—is linked to the pressure on the territory due to excessive land use in concentrated areas. Moreover, it is also due to deforestation and natural resources exploitation for the energy supply (UN 2007), representing a threat to ecosystems.

If the way to build in small villages or small settlements still reflects a link between the environment, climate and local materials, in urbanized contexts that are rapidly transforming, this connection is missed. The bioclimatic-environmental approach to design should become a common practice since it is based on local climatic conditions, assuming them to be the fundamental parameter influencing the design choices.

According to this perspective, a broad-spectrum knowledge path has to precede the design process to allow interaction with the place resources and characteristics.

In particular, as regards Mozambique—a country characterized by a long coastal strip that gradually rises towards the inlands until the edges of the great continental plateaus—it is important first to investigate the characteristics of the climate.

According to the Koppen classification, the climate in Mozambique, influenced by the monsoons of the Indian Ocean and the warm current of the Mozambique Channel, is tropical hot (sub-type Tropical Savanna Climate), varying, depending on the region, between dry climate sub-humid and semi-arid. However, it also has several regional variations due to local factors such as altitude, proximity to the coast and latitude.

The northern region is subject to low equatorial pressures, while the south is affected by tropical anticyclones and hot currents in the Mozambique Channel. More precisely, three climatic zones can be distinguished in the territory:

1. North and Centre: monsoon climate, with a dry season of 4–6 months.
2. South: drier climate, with a dry season of 6–9 months.
3. Mountainous areas: tropical climate due to the altitude.

During the year, there are two seasons: the dry and cool season, between April and October, and the hot-humid rainy season, from October to March. From October, the rains begin to intensify until March/April (only in the south, the onset of rains is often postponed) (Fig. 1).

Average annual temperatures vary between 20 °C in the south and 26 °C in the north, with the highest values during the rainy season. The average yearly humidity values (RH%) are, in general, relatively high, ranging between 65% (dry season) and 75% (hot and humid season). During the rainy season, the prevailing winds in the northern part of the State are from the northeast and in the country's southern part from the south.

The effects of climate change have determined in the temperate climate areas advantages in energy consumption and thermal comfort in the winter season (with a reduction in heating needs), despite an increase in cooling requirements. In a tropical climate, the need to minimize heat gains and maximize heat losses represents an increasing challenge throughout the year.

3   Bioclimatic Strategies to Design at the Urban and Neighbourhood Scale

Rapid urbanization, especially in Africa, generates critical issues from the environmental point of view that can be limited only through a conscious and a cross-scalar approach.

A single building does not change the climatic conditions of a city. The ensembles of buildings built according to the international standards, which generally disregard the environmental characteristics of the site and replicate similar models worldwide, are responsible for the well-known image of scarcely sustainable contemporary large cities and megacities. Cities that are nowadays responsible for 70% of greenhouse gas emissions. For this reason, it is necessary to restore environmental conditions that could improve the health of citizens and urban liveability.

Before the city scale, it would be opportune to focus on the neighbourhood one. Operating at this scale is suitable to cope with the issues that influence the environmental quality of urban areas: green and blue infrastructures, water management (from recovery to the reduction of runoff, to the use of water as cooling strategies) and mobility. Moreover, the distance between buildings for solar access and natural ventilation has to be defined at a neighbourhood scale, not at the city scale.

A well-ventilated elevation, or a river, or the breeze to intercept when settlements are built, become recognizable elements of the places, that mark a connection with the natural elements, nowadays often neglected. The prevalent multi-storey buildings which block the natural ventilation, with large openings facing the solar radiation for many hours per day, and materials with little heat capacity unable to store solar radiation, could not be used in cities with a tropical climate. Together with this type of development, the informal settlements, developed from the spontaneous aggregation of single-family buildings, do not have a link with the environmental characteristics of the urbanized site, because they arose close to an already urbanized area or to complement it. They often arise without formalizing property rights and represent agglomerations lacking in services, such as the water and sewage network, formalized public spaces, and vegetation. In the case of excessive rain and heat, this condition amplifies the inconveniences for residents and worsens the functioning of urban infrastructures: slowing down/impeding the flow of water or preventing the generation of cool niches along a path, to remind some potential problems.

Intervening today in specific highly urbanized contexts could also represent a stimulus in the generation of new nuclei capable of encouraging different types of urban development that help reconstructing a link with the environmental variables of a specific site. For example, although the general characteristics of the climate of the different areas of Mozambique are known, it is necessary to analyse the characteristics of the specific context to intervene on it. These characteristics are responsible for modifying the energy flows and the relationships between the various elements.

Strategies leading to the improvement of environmental conditions in hot-humid climates generally requires not only the control of short-wave but also long-wave (emitted heat) solar radiation, which in urban areas is connected to the morphological configuration, to the orientation of the streets according to the solar radiation and prevailing winds, but also to the presence of shading systems, vegetation and water. Strategies for improving outdoor thermal comfort (i.e. mitigating the Urban Heat Island—UHI) in a tropical climate are related to:

- The development of urban morphology (geometry of the neighbourhood, building volume) to minimize radiation trapping and enhance shadowing;
- The development of a street layout and building shape to favour wind access (Fig. 2);

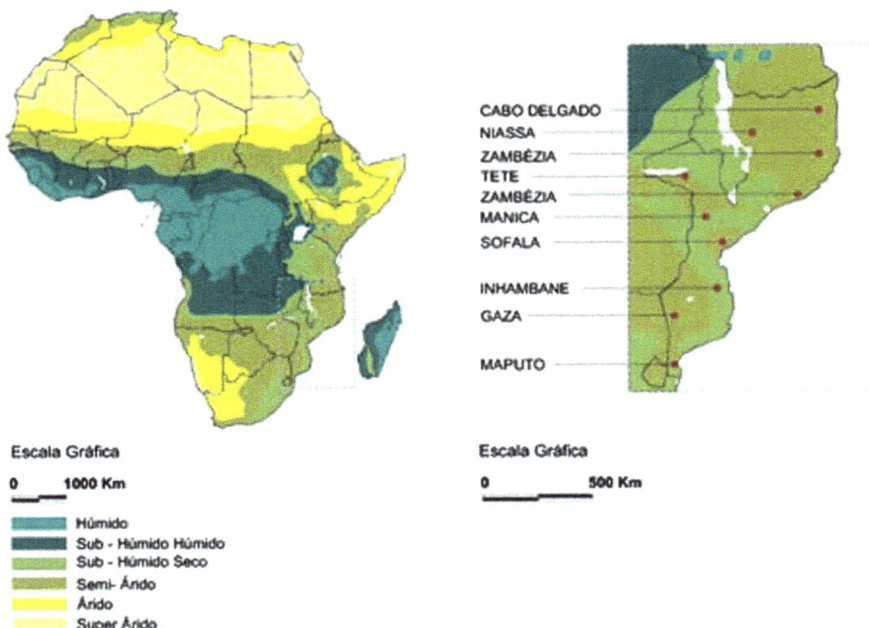

Escala Gráfica

0   1000 Km

- Húmido
- Sub - Húmido Húmido
- Sub - Húmido Seco
- Semi- Árido
- Árido
- Super Árido

CABO DELGADO
NIASSA
ZAMBÉZIA
TETE
ZAMBÉZIA
MANICA
SOFALA

INHAMBANE

GAZA

MAPUTO

Escala Gráfica

0   500 Km

Fig. 1 Distribution of aridity zones in Africa and Mozambique (according to the World Meteorological Organization—WMO)

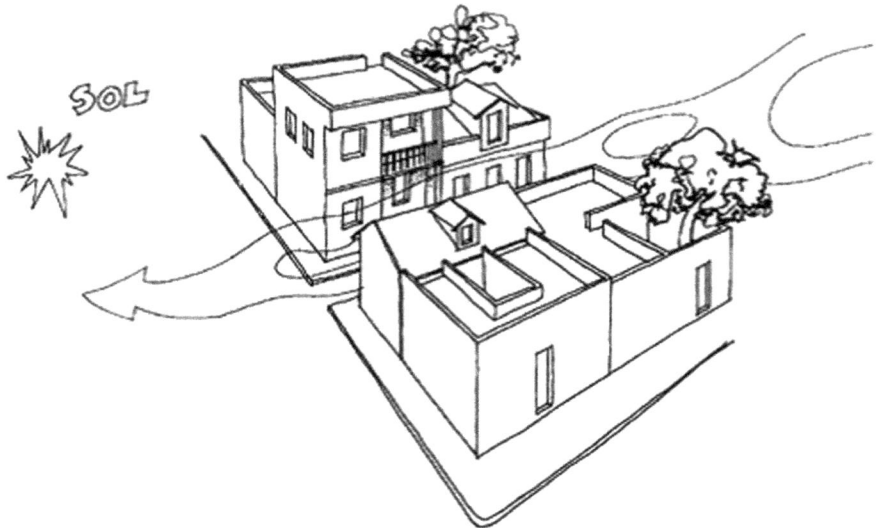

Fig. 2 Urban ventilation plays an important role both for well-being in open spaces and inside buildings, for this reason it is important to orient the streets in the direction of prevailing winds or breezes

– The control of the physical properties of urban surfaces, i.e. colour (Albedo) and thermal mass.

Increase of vegetation and—when relative humidity conditions allow—water bodies to maximize evapotranspiration loss.

As suggested by Butera (2018), for East African Community with a tropical climate (primarily if located in a latitude from about 11 °S to 5 °N), confirmed by obstruction profiles and sun charts of different types of streets, as represented in Fig. 3:

– A North–South orientation is the best. A point at the centre of the street canyon, ground level, is subject to direct sunshine for the minimum number of hours in all months; thus, better outdoor and indoor comfort is achieved, provided that the windows and walls of the upper floors are appropriately shaded with movable devices.
– With the East–West orientation, the floor of the canyon is fully exposed to solar radiation for most of the time, irrespective of the aspect ratio (H/D): only in the coolest and in the hottest month of the year the floor of the canyon will be fully shadowed for H/W $\geq$ 2. Comfort condition is complicated to reach only considering the geometry. Shading devices, vegetation and water are strongly recommended.

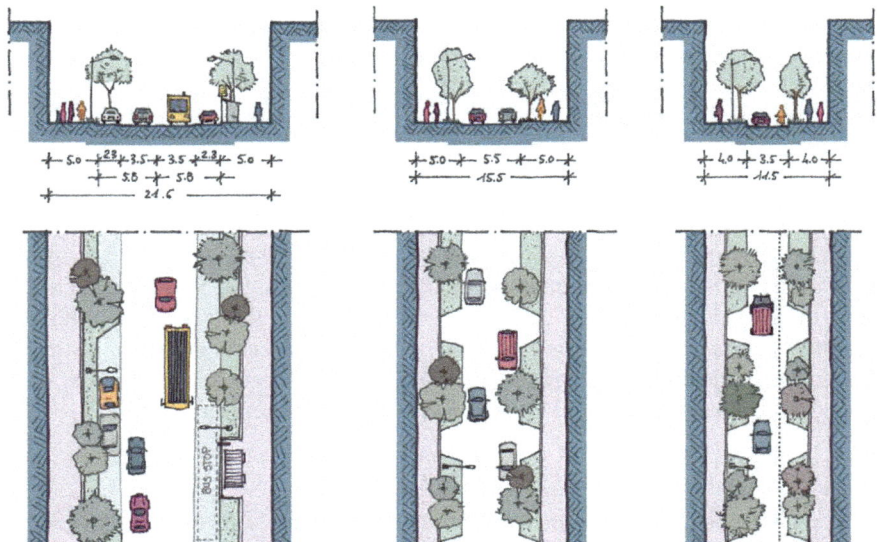

Fig. 3 Configurations of the streets according to the dimension. Left: neighbourhood connector; these streets service and link neighbourhoods. Centre: access street; to accommodate shared pedestrian, bike and vehicular movements. Right: footpath or shared path; pedestrian and cyclist street allowing temporary access to vehicles (*Credits* Butera 2018)

Urban materials and morphology determine the urban heat island, and—especially in hot climates—the albedo (linked to colour and roughness), the specific heat and density have to be taken into account.

High values of these last two characteristics guarantee the contribution of the thermal mass, i.e. the possibility to store the heat during the day and to release it in the evening, from the buildings and from the urban canopy layer (i.e. the layer of the atmosphere above the building's roof).

Massive materials (bricks, concretes, raw earth ...), colours of the cladding of buildings and non-dark urban flooring are suitable to avoid heat absorption. Moreover, as far as urban flooring is concerned, it is also essential to guarantee a good percentage of pervious materials (soil, mixed flooring, partly sealed pavements and soil).

Vegetation (in particular, trees) has multiple benefits concerning both the mitigation of the causes of climate change (absorption of CO_2) and adaptation to the climate changes, linked to the reduction of the urban heat island, which also represents an improvement of thermal comfort and a contribution to the management of excess rainwater. Trees represent a contribution at least for four reasons:

– shadow on people;
– shadow on urban surfaces, which reduces the risk of surfaces overheating;
– reduction of the air temperature values;

– reduction of the mean radiant temperature (parameter considered in calculating the thermal balance, i.e. thermal comfort).

However, it is important to emphasize that the increase of vegetation is possible if the water resource is adequate. In tropical climates with a concentration of rainwater in the rainy season, it is recommended to implement water collection systems not only at the building level but also at the urban space level to have available water to irrigate the vegetation.

The generation of new urban spaces and the renovation of existing ones in consolidated urban fabrics is fundamental for the connection of natural elements, in particular greenery and water. They allow for walking and staying in areas protected from intense solar radiation and control the flooding and excess rainwater on streets, directed on predetermined routes and not in the areas occupied by buildings. A balance between permeable and impermeable surfaces is therefore important.

4   Bioclimatic Strategies for Architectural Design

Outdoor comfort, which is strictly affected by the local climate, also determines indoor comfort, which, in turn, is the driver of the energy consumption deriving from the need for heating or cooling.

Before focusing on the building and the systems to optimize indoor thermal comfort conditions, it is important to evaluate the connection with the surrounding area and, whenever possible, the elements concerning the choice of the place, the orientation and the shape of the building, because they affect the energy flows exchange between the environment and the inhabited space.

As described above, the orientation of the streets and the buildings are important in urban areas. Orientation is referred to solar radiation exposure and, especially in hot humid climates, to prevailing winds. Even in a warm climate as in the Mozambican one, it is fundamental for the development of the houses to consider the wind regime, adequate ventilation and the consequent improvement of comfort indoor, as developed in the project of Condominio do Caracol in Maputo designed by Josè Forjaz (Fig. 4). Optimizing the orientation is possible to avoid overheating situations, which is the first step towards promoting heat protection and dissipation strategies.

For the Mozambican territory, the best orientation for the main facade is the North, with a deviation that must not exceed 45° from the north.

With this in mind, the sizing of the glazed areas must be compatible with the orientation of the facade; the kitchen space must be the coolest in the house, having its heat production, and therefore cannot be oriented to the west. The best orientation to reduce heat gains from solar radiation will be the development along the East–West axis, limiting the exposure area of the east and west facades.

When the solar radiation reaches the north-facing facade, it reaches a very high angle and can be easily shaded while ensuring good lighting. In urban situations where the choice of orientation usually is impossible, it can be compensated by

Fig. 4 Condominio do Caracol at Maputo. A residential building that include different strategies to limit overheating indoor. In particular, the most important strategies implemented are: Integration of planted areas on all exterior surfaces of the houses, an adequate orientation, the cross-ventilation, the shading device for the openings (*Credits* Forjaz 2011)

strengthening other strategies suitable for controlling solar gains, such as the shading of walls and windows and their sizing (Fig. 5).

In terms of building shape, the configuration and layout of the interior spaces—depending on the function—influence the exposure to incident solar radiation and the availability of ventilation and natural lighting. A compact building will generally have a relatively small exterior wall, i.e. a low surface/volume ratio. For small- and medium-sized buildings, this situation can offer advantages because the heat exchanges with the surrounding are limited, but, at the same time, it could be limited and the possibility to have cross-ventilation.

A fundamental role is played by materials with high thermal mass, as already mentioned, with the possibility to store heat during the hot hours of the day. Depending on the type of material and the thickness of the element, the release can be after hours, in particular in the evening, when the outdoor temperature begins to drop. They can be external walls, internal partitions, flat roofs or a green roof (Fig. 6). Also in traditional architecture, currently used out of the cities, as in the example of the school in Capo Delgado (Fig. 7), is possible to implement strategies that improve thermal performance of the buildings.

Direct solar radiation is by far the primary energy source that heats the building. For this reason, cooling strategies must be accompanied and anticipated by measures that reduce the risk of overheating, as in the school designed in 2010 by the architects Ziegert Roswag Seiler, with the aim to preserve and improve the local architectural tradition, using renewable resources such as earth and bamboo, managing to provide

Fig. 5 Abreu, Santos and Rocha building in Maputo (1953–1956) designed by the architect Pancho Guedes. The shading of the loggias of the individual lodgings is resolved through a composition of brise-soleil which also allows ventilation (*Credits* Forjaz 2011)

Fig. 6 The graph compares air temperature and the surface temperature of material with high thermal mass. It is possible to observe that the use of massive material helps to reduce indoor air temperature and to shift the peak hours later, according to the material and thickness Credits M. Grosso

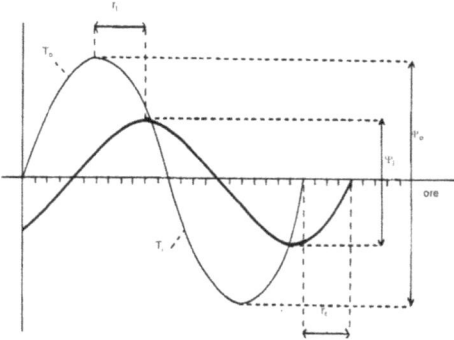

a good internal temperature and regulate humidity (Fig. 7). First, it is appropriate to define systems that prevent solar radiation from reaching the interior of the building, in particular:

– large projecting roofs for shading the external walls and windows,
– horizontal protrusions (along the north side) and vertical (on the east and west sides),
– light colour for the facades to reflect the radiation before it passes through the outer wall,
– light roofs to allow micro-ventilation of the roof structure and

Fig. 7 A series of strategies combined in a single building, a school in Capo Delgado, in the northern part of Mozambique: massive masonry, a large roof to protect masonry and openings from solar radiation, large openings for natural ventilation, the presence of arcades for carry out part of the activities outdoors, light cover to allow micro-ventilation of the structure (*Credits* ZRS Architekten Ingenieure, *Photo* Paula Holtz)

– alternatively, massive structures and roofs, such as green roofs and green walls to increase the heat storage.

In any case, as far as possible, the space around the building should also be involved in the control of the internal microclimate. The use of vegetation, water and other shielding elements can help a lot, regardless of the strategies implemented in the envelope and the elements that configure the building.

Strategies aimed at dissipating the heat produced inside the building (free earnings from electrical equipment, people, stoves…) are based on what is called "thermal wells". These techniques avoid overheating, bringing the internal temperature values to levels near to or even lower than the external air temperature (Fig. 8).

The best design solutions for passive cooling combine different strategies to achieve greater efficiency, such as heat dissipation through night ventilation of the heat re-emitted by the massive element (external wall, roof, floor).

The effectiveness of passive cooling techniques can often be improved through the use of systems that work with energy from renewable sources, such as solar or photovoltaic panels, or systems with low energy consumption from non-renewable sources, such as fans (Fig. 9).

In the Mozambican climate context, it is possible to balance the building and the climate by applying a series of design strategies deriving from the bioclimatic approach, well known in the vernacular and popular building systems, which can also be reinterpreted for contemporary architecture. Today, these traditional techniques are enhanced with the technological knowledge currently available and optimized so that they can be successfully incorporated into the design and operation of buildings. These strategies and techniques allow buildings to adapt to the surrounding environment through architectural design and the conscious use of materials and

Air that allows the natural ventilation (cross ventilation and from chimney effect)	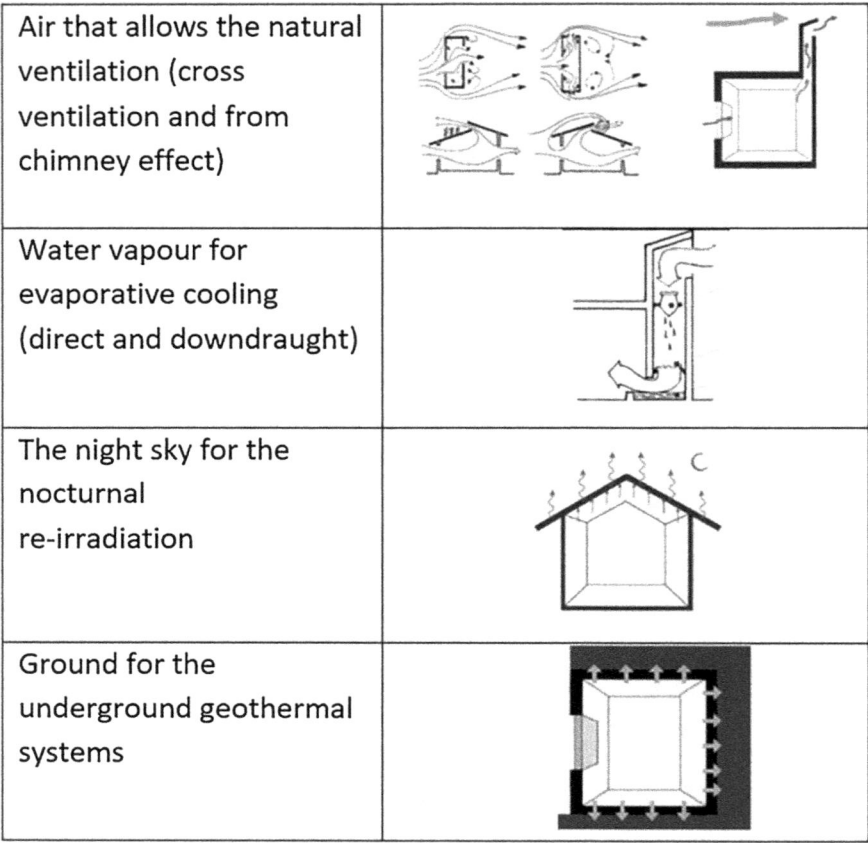
Water vapour for evaporative cooling (direct and downdraught)	
The night sky for the nocturnal re-irradiation	
Ground for the underground geothermal systems	

Fig. 8 A set of cooling strategies based on natural "thermal wells": air, water vapor, nocturnal sky, ground Credits M. Grosso

construction elements, avoiding (passive architecture) mechanical systems based on energy, often derived from non-renewable sources.

The existing building, that can be placed within a consolidated and/or compact urban fabric, offers reduced possibilities to react to the excesses of the climate compared to the design from scratch of an artefact born from the correct interpretation and enhancement of environmental, climatic and site parameters. Intervening on existing buildings, even if limited to a few elements, always depends on the type of building and on the social context. Therefore, it is essential to activate participation processes with the inhabitants who can identify the potential for improvements. A single-family building in an informal context could be renovated by modifying the envelope, increasing the roof surface to improve shading both inside the building and the external walls of the space around the building. The presence of plants could

Fig. 9 The combination of cooling strategies that increases the effectiveness. In this case the thermal mass is associated with the natural ventilation that removes the heat from the building surfaces and moves it outside from the openings Credits M. Grosso

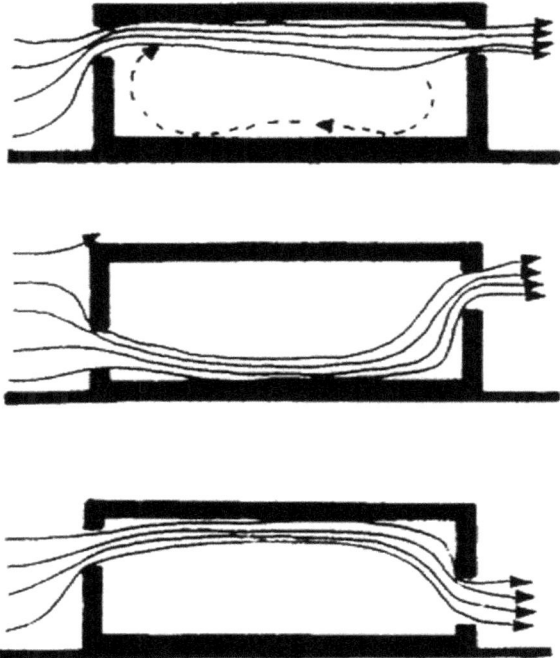

shade the roof and the external areas around the house and the possibility of collecting and reusing rainwater that could also be used for irrigation.

A multi-storey residential or office building can be modified, especially in terms of envelope and roof. A ventilated façade on an existing envelope could provide micro-ventilation and shading of the wall. Shielding systems of the openings to the outside could guarantee the passage of air and shield the solar radiation, along with mobile screens, or part of the facade itself, as in the already mentioned case of the screening of the loggias in the Abreu, Santos and Rocha building in Maputo (Fig. 4).

References

Butera F et al (2018) Energy and resource efficient urban neighbourhood design principles for tropical Countries. Practitioner's guidebook. UN-Habitat, Nairobi

Chiesa G (ed) (2021) Bioclimatic approaches in urban and buildings design. Springer, Cham

Crutzen PJ, Stoermer EF (2000) The 'Anthropocene'. IGBP Newslett 41:17–18. http://www.igbp.net/download/18.316f18321323470177580001401/1376383088452/NL41.pdf. Accessed on 10 December 2021

Forjaz J (ed) (2011) Arquitectura Sustentàvel em Mocambique. Manual de boas pràticas. Comunidade dos Países de Língua Portuguesa (CPLP), Lisbon

Gabaglio R et al (2020) Sustainability in developing countries. An approach for an enhancement of heritage in the coastal area of Mozambico. In: Amoeda R, Lira S, Pinheiro C (eds) Heritage

2020. Proceedings of the 7th international conference on heritage and sustainable development. Green lines institute for sustainable development. Barcelos, Portugal, pp 343–352

Grosso M (2008) Il raffrescamento passive degli edifici. Maggioli (ed) Sant'Arcangelo di Romagna (Rn)

IPCC (2014) Climate change 2014—mitigation of climate change—working group III contribution to the fifth assessment report of the intergovernmental panel on climate change, Chapter 12. Available via IPCC. https://www.ipcc.ch/report/ar5/wg3/. Accessed 10 December 2021

Olgyay V (1963) Design with climate. Princeton University Press, Princeton, Bioclimatic approach to architectural regionalism

Oke RR et al (2019) Urban climates. Cambridge University Press, Cambridge

UN—United Nations (2007) The Millennium development goals report 2007. UN, New York

UN—United Nations (2017) Economic report on Africa 2017: urbanization and industrialization for Africa's transformation. UN, Addis Abeba

WMO—World Meteorological Organisation (2021). https://public.wmo.int/en. Accessed on 10 December 2021

Urban Forestry: Perspectives from Sub-Saharan Africa Between Planning and Global Challenges

Maria Chiara Pastore

Abstract Today, Urban Forestry and, more generally, the actions of planting and maintaining forests are at the forefront of the International Agendas. As the pressure towards the climate change effects arises, urban forestry and, more generally, urban greening become an increasingly critical infrastructure and an essential public service (Konijnendijk in Nature-based solutions for more sustainable cities. Emerald Publishing, Bingley, 2021). The essay aims at providing an introductory discussion on the specific challenges related to urban forestry in sub-Saharan Africa, and more specifically to Mozambique, to provide the context of the research to date in the sector of urban forestry in sub-Saharan Africa, and to introduce the discussion on the specific context of the metropolitan area of Maputo, where the pace of development and urbanization, in a changing climate condition, are rapidly deteriorating the overall ecosystem.

1 Introduction

Today, Urban Forestry and, more generally, the actions of planting and maintaining forests are at the forefront of the international agendas.

Worldwide, there has been a constant and growing concern on how to improve the quality of life in our cities, making them 'healthier, happier, greener cooler, wilder' (FAO 2018). The renovated effort on greening our city comes in response to the grand challenges faced by human society in our time, such as climate change, biodiversity loss, increased inequality and, more recently, the COVID-19 pandemic.

As the pressure towards the climate change effects arises, urban forestry and, more generally, urban greening become an increasingly critical infrastructure and a basic public service (Konijnendijk 2021). In this context, the term "urban forestry", firstly mentioned in 1894 (Cook 1894), has been experiencing a renewed interest since the 1960s with the gradual recognition of its potential and substantial role in making

M. C. Pastore (✉)
Dipartimento di Architettura e Studi Urbani, Politecnico di Milano, Milan, Italy
e-mail: mariachiara.pastore@polimi.it

cities more liveable and sustainable in the long term. Nowadays, urban forestry can be defined as 'an integrated concept, defined as the art, science, and technology of managing trees and forest resources in and around community ecosystems for the psychological, sociological, aesthetic economic, and environmental benefits trees provide society' (Konijnendijk et al. 2004: 472). To provide a more specific context, an urban forest is defined as 'comprising all tree-dominated green areas in and around urban areas' (Konijnendijk et al. 2004: 472). It includes, according to FAO, forests, other wooded lands and trees outside forests, in urban environments.

It is necessary, in order to discuss the topic of urban forestry in the context of sub-Saharan Africa and in Maputo and Mozambique, to provide an introductory discussion on the challenges referring to urban forestry in sub-Saharan Africa, which is part one of this chapter. The second part focuses on Mozambique and Maputo by presenting the specific challenges and opening the perspective on the subject.

2 Sub-Saharan Africa Context and Urban Forestry

In order to introduce the subject, I would like to introduce two arguments.

First, it is difficult to talk about urban forestry and, in general, urban green systems because the relationship between green systems and the African city are not very treated from a scientific point of view. Although there is much international research on urban forestry issues, contributions from developing world countries in the peer-reviewed literature and seminal texts seem limited (Shackleton 2012).

In her paper, Hosek analysed peer-reviewed, English language literature on the benefits and services of urban trees in African countries to assess which is the state of research on urban forestry in the African continent and to identify issues that have received limited attention and possibly the perspectives (Hosek 2014). Out of 44 publications included in the literature review, only ten countries appear to tackle the subject, but considering the year of publication, it is interesting to point out that there is a growing interest in the subject, as the first publication dated 1988, and only in the year 2013, there is a total of ten publication on the subject.

Another paper by Collins Adjei Mensah shows that, giving the general lack of research on the subject in the African context, the few studies are not broad base, but they mostly concentrate on green spaces within a particular city (Mensah 2014).

A third study investigates (Wangai et al. 2016) the Ecosystem Service research in Africa, focusing on spatial distribution, criteria and methodologies used in the studies. From the open search that resulted in 709 scientific articles, 52 were considered for the review. Results indicate that most studies were conducted in South Africa, Kenya and Tanzania, and focused on services provided by watersheds and catchment ecosystems. Most of the studies regard the provisioning of multiple ecosystem services, particularly economic valuation and mapping (Wangai et al. 2016).

If the research shows limited but growing interest in the research related to urban forestry, green systems, ecosystem services, so do the planning and development processes in expanding cities (Du Toit et al. 2018; Mensah 2014). However, it is important to point out that the provision of green systems is seminal in African cities, as the high level of inequality towards access to basic services and great level of urban poverty demand reliance on green systems for the provision of essential ecosystem services, such as water, fuel and food production (Anderson et al. 2013).

This consideration leads to the second point of the discussion that it is extremely relevant to talk about the relationship between cities and greening systems because Africa is urban, green systems are shrinking and unequal in terms of access and ownership.

The impact that the rapid increase of the urban population is having and will have in sub-Saharan Africa (United Nations 2015) is a severe threat to the permeable and green systems on a local and global scale (Güneralp et al. 2013; Seto et al. 2012).

Current African ecosystem degradation (AEO 2013) can be compared with what occurred during the Industrial Revolution of the nineteenth century in Europe (Wangai et al. 2016; Gafta and Akeroyd 2006). In particular, the vulnerability to climate change and desertification are expected to escalate due to human malpractices such as deforestation (IPCC 1997).

As mentioned by Mensah, studies on several African countries revealed that 'there is intense pressure on green spaces for different human activities resulting in persistent deterioration of these spaces especially in urban areas where the pressure is more profound' (Mensah 2014: 1). Now, the rapid depletion of green spaces in Africa has resulted in green spaces occupying a very small per cent of the total land space of many urban areas.

If the green spaces are deteriorating and shrinking, their accessibility by the population living in urban spaces is also declining. Worldwide, there is a growing concern about the inequitable patterns of green systems in urban areas, and the potential predictability increases inequality of access by areas or populations (Venter et al. 2020). It is demonstrated that parks and recreational spaces are less available in low socio-economic status and high minorities communities (Hughey et al. 2016). Again, the literature that discusses the disparities in access to green spaces are limited in sub-Saharan Africa, but the recent research conducted by Venter in South Africa provides specific insights into the specific challenges (Venter et al. 2020). The paper highlights that the provision of green spaces in South Africa is still very much linked to the Apartheid city, 'designed around the spatial segregation of these race groups, with people forcibly removed to 'group areas' and regulations around social interaction in public space' (Venter et al. 2020), and even if the laws were officially dismantled in 1994, the legacy of the planning choices made in the past 50 years are still in place, particularly in the provision of all the basic infrastructure, including the green spaces. The same element of poor provision of essential services (and green spaces) during the colonialism period in many African cities reflects the lack of institutionalized green spaces in the contemporary African city. In addition to that, the incredible pace of urbanization, either formal or informal, creates the perfect threat to those spaces that host green or blue systems.

This second point allows me to frame some of the most pressing challenges very specific to urban forestry in sub-Saharan Africa.

The first, already mentioned, relates specifically to the extraordinary pace of urbanization happening in the sub-Saharan cities, where informal and formal settlements are taking place often at the expense of green systems.

More generally, when we speak of urban forestry, there is not only a land use competition between those green spaces that host trees and the built environment, but the situation is far more complex. The competition is also in those spaces that provide profits, for instance, the agriculture sector. Deforestation continues to be driven by the need to produce food. In 2018, as stated by FAO, the most significant loss of forests has been in tropical and low-income countries, which have also experienced the most remarkable expansion of agricultural land. It is estimated that 80% of forest loss is due to conversion to agriculture. Uncontrolled fires, usually by human activity and in strict relation to agriculture, are one of the main causes of forestry loss.

Another crucial element is related to the value of wood. Among the different elements that make wood valuable, I would recall two elements specifically: the first relates to commercial logging, particularly for timber or pulp (construction sector and paper products), the other refers to charcoal production. While commercial logging is more related to vast land areas, in urban areas, trees are often used as material for construction, particularly in informal settlements, being a partial cause for deforestation. In reference to charcoal production, charcoal represents 9% of the total primary energy supply worldwide and 27% of the primary energy supply in Africa. An estimated 2.4 billion people depend on wood energy for cooking and/or heating (FAO 2018). This means that formally and informally, trees are still significantly endangered as a source of energy production.

A fourth element is, of course, climate change, which is intensified by the loss of forests. Periods of intense drought, flash floods, fires, insects' invasions, variations in temperatures are all harmful to the green systems, which is already suffering degradation because of the urbanization processes.

Another element that I would like to raise is more related to the competition over the provision of services. Green infrastructure, and generally the protection of existing green systems, are not prioritised in planning and development processes because of costs but also because of the complex nature of urban planning regulations and the bureaucracy involved in issuing permits within the planning systems. This often allows developers and other individuals to evade the required planning procedures and to embark on projects which do not put green systems as part of the development (Mensah 2014).

All these elements, which characterise the condition of urban forestry in sub-Saharan Africa, are all present in the case study analysed, the Greater Maputo, and in general in Mozambique.

3   On Urban Forestry in Maputo, Mozambique

Mozambique has 32 million hectares of natural forests, covering 40% of its area (World Bank 2018b). Of the 32 million ha, 27 million ha are categorized as production forests (Unique 2016). The main forest ecosystem is the *miombo*, a vast region of tropical grasslands, savannas and shrublands, while other forest ecosystem types include the coastal forests in the south, afro-montane forests in central Mozambique, and coastal dry forests in the north. Mozambique also hosts the second largest area of mangroves in Africa (World Bank 2018b). The agricultural sector employs more than 80% of the Mozambican population and accounts for 32% of the country's GDP (Armand et al. 2019).

Mirroring the discussion above presented, forests are heavily threatened within the country. Every year 267.000 ha of forests are lost, 65% of it due to slash and burn agriculture, 12% of urban expansion, 8% for timber production, 7% for charcoal and 8% for other reasons (World Bank 2018a).

The government of Mozambique, recognizing the importance that forests play within the national economy, mainly in rural areas, representing a source of energy (firewood, charcoal), construction material, non-timber forest products and nutrients for small-scale agriculture (Chidumayo and Gumbo 2010; GoM 2018), embarked on the development of the country's Forest Sector Agenda 2035.

The three main strategic assets (1) Forests for socio-economic development and food security with a focus on the involvement of local communities; (2) Forests in building resilience to climate change and natural disasters; and (3) Governance and capacity building; mentioned within the 'Agenda Estrategica 2019–2035 e programa nacional de florestas' (MITADER 2019: 9) help us to recognize the main challenges that the country needs to face within a relatively short term.

As shown in Fig. 1, deforestation affects provinces unequally, representing a risk for the perpetuation of forest heritage in those provinces that combine higher population density and less forest resource, such as Nampula (predicted loss of 58% of current forests by 2035), Sofala (27%) and Manica province (24%) (MITADER 2019). Based on statistics available for the countries, the study estimates that districts in the coastal region of the north-central region and the Beira and Nacala corridors will be the most affected by deforestation, namely, the districts of Gondola, Sussundenga, Macate, Lalaua, Mogincual, Moma, Liupo, Meconta, Monapo, Nacala-a-velha, Buzi, Chibabava, Nhamatanda and Alto Molocué.

In this context, it is interesting to review the case of the Maputo Province, where the two cities of Maputo and Matola only, with a combined population of more than 2.6 million inhabitants, represent 30% of the country's urban population and the largest Mozambican urban area (de Araújo 2006) (Boa_Ma_Nhã, Maputo! 2020). All districts in the Maputo Province have experienced growth in the period 2007–2017. Matola and Boane districts have seen an extraordinary demographic boost doubling their inhabitants, while others like the Namaacha district have experienced a positive but less outstanding population increase (Boa_Ma_Nhã, Maputo! 2020: 43).

As reviewed by the Research Project "Boa_Ma_Nhã, Maputo!" a complex nexus of water, energy, food defines both the challenges and the possible solution towards a more sustainable pattern of development.

In the case of the city of Maputo and its province, I would like to point out a few elements that can possibly help the discussion on the subject. The first concerns the

Porporção de perda de área florestal (%)   *Quantidade de área florestal perdida (ha)*

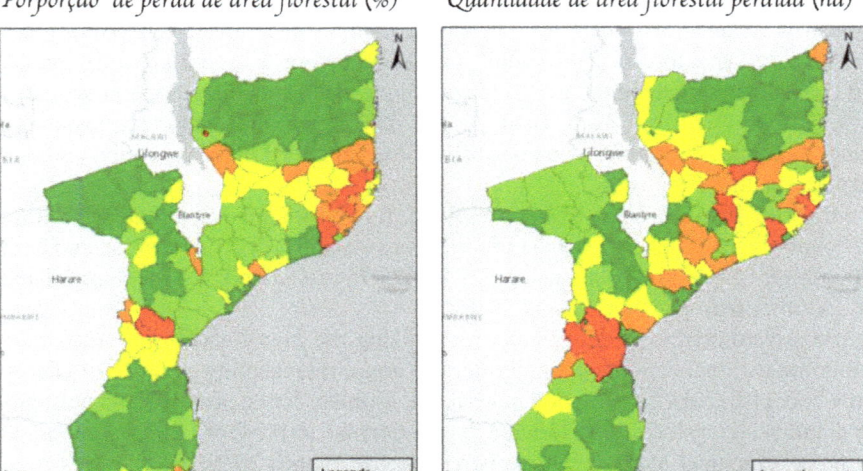

Fonte: Mabilana, 2019

Fig. 1 Vulnerable districts due to loss of proportion of forest area (>40%) and the amount of area to be lost (>60,000 ha) (MITADER 2019: 48)

complex ecosystem, which as Vasco and Costa point out, is one of the most diversified of Africa and falls within the Savanna biome being a mosaic of sand forests, scrub forests, evergreen and semi-evergreen bush-land and thicket, in a matrix with secondary and wooded grasslands. Mangroves, areas still dedicated to agriculture within the city, in a dimension of complicated coexistence with urban expansion, such as Catembe, and Costa do Sol, Marracuene, very dense and consolidated residential areas such as Maxaquene, Mafalala, the infrastructure of the airport now at the centre of a city in continuous expansion, and then, again, neighbourhoods that reflect a vocation linked to agriculture, linked to the Rio Infulene, and again, residential neighbourhoods that are rolled down up to the Rio Matola, in a continuous succession of dense inhabited areas which now follow the orography, now the main infrastructures.

Those green spaces in the city, which are particularly evident from far, represent either complex green environment (the rivers, the marshland areas), some floodplain areas that resist the urbanization process, and bounded elements such as military areas and schools, private areas. Walking in the city, a different grain of green spaces appears, often behind a fence, a brick wall or a corrugated iron sheet used to delimitate a private space. Shrubs are also part of this domestic landscape of fences, particularly used in peri-urban spaces, appearing in those neglected spaces such as small informal garbage sites, accompanying those infrastructures that would benefit some maintenance (i.e. stormwater drainage). Few tree-lined avenues appear in the central business districts.

Tree products, on the contrary, are particularly evident in all those markets made of wood and wooden products, the selling of charcoal, the selling of bamboo and other wood construction material, it is there in all the fruits appearing in the markets, and all those agriculture products.

It is then evident that the three districts, Boane, Namaacha and Moamba, are providing crucial services for the survival and growth of Maputo in terms of water, energy and food provision. In addition, at the moment, those three districts appear to provide the only future land reserve to host all those ecosystems services that are vital for sustaining the metropolis. This interdependence should be fully recognized and re-balanced in defining the effective consistency of Maputo's metropolitan area, territorial footprint and related governance tools.

4 Conclusive Remarks

This contribution aims at providing a context to the term urban forestry, which, in sub-Saharan Africa relates to specific challenges such as deforestation, competition on the different land uses, environmental degradation, and wildfires. As globally and locally, planting new trees is seminal to the wellbeing of the overall environment, it is necessary to build a road map where the integration among the different development strategies and the green systems become structurally embedded. Moreover, recognizing that urban forestry does not belong to a specific sector, being related to conservation of the environment, agriculture sector, energy, green infrastructure, food strategies, to mention a few, but also pertaining to different scales, either local, city, district, regional and national level, it is necessary to continuously provide relations, through the planning processes, of the different design processes and the strategic level of connectivity that trees and plant provide to our systems.

Acknowledgements This chapter has been prepared with the support of Giulia Pregagnoli.

References

AEO—Africa Environment Outlook (2013) Africa Environment Outlook 3: summary for policy makers. A publication of the United Nations Environment Program. Available via UNEP. https://wedocs.unep.org/handle/20.500.11822/8653. Accessed 15 Dec 2021

Anderson PL, Okereke C, Rudd A, Parnell S (2013) Regional assessment of Africa. In: Elmqvist T, Fragkias M, Goodness J et al (eds) Urbanization, biodiversity and ecosystem services: challenges and opportunities a global assessment. Springer, Dordrecht, pp 453–459

Armand A, Gomes JF, Taveras IK (2019) Managing agricultural risk in Mozambique. International Growth Centre, London

Boa_Ma_Nhã, Maputo! (2020) Polisocial development plan. Politecnico di Milano, Milan

Chidumayo EN, Gumbo DJ (2010) The dry forests and woodlands of Africa: managing for products and services. Routledge, London

Cook GR (1894) Report of the general superintendent of parks. In: Second annual report of the Board of Park Commissioners, Cambridge, Massachusetts, pp 71–98

de Araújo MGM (2006) Espaço urbano demograficamente multifacetado: As cidades de Maputo e da Matola. Available via APDemografia: http://apdemografia.pt/files/1853187958.pdf. Accessed 15 Nov 2021

Du Toit MJ et al (2018) Urban green infrastructure and ecosystem services in sub-Saharan Africa. Landsc Urban Plan 180:249–261. https://doi.org/10.1016/j.landurbplan.2018.06.001

FAO (2018) World forum urban forests 2018 greener, healthier and happier cities for all. A call to action. FAO, Rome

Gafta D, Akeroyd JR (eds) (2006) Nature conservation: concepts and practice. Springer Science & Business Media, New York

GoM - Governo de Moçambique (2018) Mozambique's Forest reference emission level for reducing emissions from deforestation in natural forests (p. 50). Available via Ministério da Terra, Ambiente e Desenvolvimento Rural. República de Moçambique. https://redd.unfccc.int/files/2018_frel_sub mission_mozambique.pdf. Accessed 15 Dec 2021

Güneralp B, McDonald R, Fragkias at al. (2013) Urbanization forecasts, effects on land use, biodiversity, and ecosystem services. In: Elmqvist T, Fragkias M, Goodness J et al (eds) Urbanization, biodiversity and ecosystem services: challenges and opportunities a global assessment. Springer, Dordrecht, pp 437–452

Hosek L (2014) Urban forestry in Africa—insights from a literature review on the benefits and services of urban trees. In: Johnston M, Percival G (eds) Urban trees research conference, 2–3 April 2014 University of Birmingham, Edgbaston, UK, pp 43–53

Hughey SM, Walsemann KM, Child S, Powers A, Reed JA, Kaczynski AT (2016) Using an environmental justice approach to examine the relationships between park availability and quality indicators, neighborhood disadvantage, and racial/ethnic composition. Landsc Urban Plan 148:159–169

IPCC—Intergovernmental Panel on Climate Change (1997) In: Watson RT, Zinyowera MC, Moss RH (eds) The regional impacts of climate change: an assessment of vulnerability. IPCC/Cambridge University Press, Cambridge

Konijnendijk C (2022) What are nature-based solutions? The potential of nature in cities. In: Croci E, Lucchitta B (2021) Nature-based solutions for more sustainable cities. Emerald Publishing, Bingley

Konijnendijk van den Bosch C, Randrup T (2004) Urban forestry. In: Burley J, Evans J, Youngquist JA (eds) Encyclopedia of forest sciences. Elsevier Sciences

Mensah CA (2014) Urban green spaces in Africa: nature and challenges. Int J Ecosyst 4(1):1–11. https://doi.org/10.5923/j.ije.20140401.01

MITADER - Governo de Moçambique - Ministério da Terra, Ambiente e Desenvolvimento Rural (2019) Agenda Florestal 2035 e Programa Nacional de Florestas Versão após harmonização multisectorial, MITADER, Maputo

Seto KC, Guneralp B, Hutyra LR (2012) Global forecasts of urban expansion to 2030 and direct impacts on biodiversity and carbon pools. In: Turner BL (ed) Proc Natl Acad Sci 109(40):16083–16088. https://doi.org/10.1073/pnas.1211658109

Shackleton CM (2012) Is there no urban forestry in the developing world? Sci Res Essays 7(40):3329–3335. https://doi.org/10.5897/SRE11.1117

UN - United Nations (2015) World urbanization prospects: the 2014 revision. Department of Economic and Social Affairs, Population Division, New York

Unique Forestry and Land Use (2016) Financial analysis of the natural forest management sector of Mozambique. Mozambique Forest Investment Project. Unique Forestry and Land Use, Germany

Vasco H, Costa M (2009) Quantification and use of forest biomass residues in Maputo province, Mozambique. Biomass Bioenergy 33(9):1221–1228. https://doi.org/10.1016/j.biombioe.2009.05.008

Venter ZS, Shackleton CM, Van Staden F, Selomane O, Masterson VA (2020) Green Apartheid: urban green infrastructure remains unequally distributed across income and race geographies in South Africa. Landsc Urban Plan 203(103889). https://doi.org/10.1016/j.landurbplan.2020. 103889

Wangai PW, Burkhard B, Müller F (2016) A review of studies on ecosystem services in Africa. Int J Sustain Built Environ 5(2):225–245. https://doi.org/10.1016/j.ijsbe.2016.08.005

World Bank (2018a) Forests of Mozambique: a snapshot. Available via the Word Bank. https://www. worldbank.org/en/news/infographic/2018a/12/12/forests-of-mozambique-a-snapshot. Accessed 15 Oct 2021

World Bank (2018b) Mozambique. Mozambique country forest note. Available via the World Bank. https://documents1.worldbank.org/curated/en/693491530168545091/pdf/Mozamb ique-Country-Forest-Note.pdf. Accessed 15 Dec 2015

Governance: Rethinking Paradigms and Urban Research Approach for Sub-Saharan African Urbanism

Paola Bellaviti

Abstract To address the enormous challenges of management and rebalancing of the massive "urbanization of poverty" widespread in many countries of the Global South and particularly in the African continent, international multi-later organizations such as the World Bank and UN-Agencies have long since introduced the notion of *governance*—reformulated as *good governance*—as a sort of "magic formula" to tame unplanned and informal urban growth and enable a more prosperous, equitable, and sustainable urban development. The practical application of this new mode of urban government that has been so successful in Western cities—although still debated—has encountered and still encounters many obstacles in contexts where the government, especially at the local level, is weak and poorly equipped, formal resources are very limited, and informal processes predominate. The essay tries to reconstruct this problematic framework, especially with reference to sub-Saharan Africa, drawing on the growing studies on the specificity of African urbanism, which strongly support the need for a "place-based innovation" of planning and urban governance based on specific knowledge production and rooted in a new theory and praxis of urban research in that context. In the end, the case of the action-research "Boa_Ma_Nhã, Maputo!" is argued as a valuable contribution to this perspective.

1 The Concept of Governance in Transition Between Global North and Global South

The African continent, due to the very rapid urbanization of the last decades, is facing severe challenges in the field of urban and territorial management, often without the adequate technical and political-administrative resources—in addition to economic ones—to provide the necessary infrastructures and manage the decision-making and regulatory processes of the "urban revolution" underway (Parnell and Oldfield 2016).

P. Bellaviti (✉)
Department of Architecture and Urban Studies, Politecnico di Milano, Milan, Italy
e-mail: paola.bellaviti@polimi.it

L. Montedoro et al. (eds.), *Territorial Development and Water-Energy-Food Nexus in the Global South*, Research for Development,
https://doi.org/10.1007/978-3-030-96538-9_15

209

Therefore, the UN-Habitat Report *The state of African Cities 2014: re-imagining sustainable urban transition* (UN-Habitat 2014) clearly indicated the need, in the face of the profound socio-spatial inequalities and the strong environmental imbalances characterizing African urban transition—tangibly represented by the proliferation of massive urban slums lacking the minimum requirements of habitability and sustainability—to overcome the models of urban development, planning, and management imported from Western countries of which African countries were colonies. The report thus highlighted the need to develop new approaches, methods, and institutional capacities for adapting urban management to the specificities of the African city in its various macro-regional variations.

To support this cultural and institutional change, in the African context as in the rest of the Global South, multi-lateral organizations such as UN-Habitat, the World Bank, etc. have long since introduced the notion of *governance* as a sort of "magic formula" to tame unplanned and informal urban growth and foster a more prosperous, equitable, and sustainable urban development.

As Obeng-Odom (2013, 2017) effectively summarizes introducing his studies on governance strategies in Africa: "governance has been used as a political and economic concept for (and certainly as a solution to) all the challenges on the continent", underlining how: "The world development institutions commonly present 'urban governance' as an antidote to the so-called 'urbanization of poverty' and 'parasitic urbanism' in Africa".

But which kind of governance can be exercised in contexts so different from those of the Global North where the concept of urban governance was born and triumphed in a phase of profound economic, political and spatial restructuring connected with globalization?[1]

Let's briefly review this epochal transformation in the history of Western urban and territorial government, before delving into the problems of its application to cities of the Global South and especially to the "rogue urbanism" of the African continent (Pieterse and Simone 2013), taking up a key text by Petrillo (2017) on the possibility of "governing the ungovernable", that is, exercising governance strategies in an urban setting dominated by informality.

Introduced and developed in the 80s/90s in Western countries in relation to the emergence of neoliberal models of government, the concept of governance marked the transition from big government, "the classic form of public administration of the post-war Welfare-Keynesian systems, which was in charge of guaranteeing services and redistributing income using a traditional rational bureaucracy, organized by hierarchies of authority", to "a less defined 'urban governability'" that is meant as "system of government that articulates and associates political institutions, social actors and private organizations, in processes of elaboration and realization of collective choices, capable of provoking an active adhesion of citizens" (Petrillo 2017: 35).[2]

This "historical transition" has deeply marked the modalities of urban management, thus delegated to a plurality of institutional and non-institutional structures,

---

[1] See Brenner (1999, 2004), Brenner and Theodore (2002), in References.

[2] The quotations from Petrillo (2017) are translated from Italian by the Author of this essay.

to public–private partnerships and networks of collective actors in which the private sector plays a decisive role, but where also bottom-up participation of citizens can play a role, thanks to the reticular-horizontal nature of the decision-making processes configured by governance. More widely, this change has supported a profound reorganization of spaces and of political and economic powers on a global scale and has relocated urban policies in this new scenario of global and multi-scalar development in terms of economic competition between cities and urban regions to attract the most valuable global functions.

The transition "from government to governance" is now a *fait accompli* in Western countries, where it has had impacts generally considered positive in terms of the effectiveness of urban management—thanks also to a relevant innovation in planning tools and practices, reformulated in a strategic and participatory way—but it has induced changes in urban societies that are very controversial and still debated, as effectively underlined by Petrillo (2017: 37–38):

> ... the repercussions if measured at the city level have been enormous and have led to a growing presence of private, a redefinition of balances and powers, a remixing of populations, a redefinition of places of life and local identities. In this sense, the debate on the forms and meaning of urban governance embraces a whole series of issues and retains many ambiguities. Although governance theoretically exercises its action horizontally, through networks of collective actors, in which subjects enter the negotiation processes through mediation and consent procedures, these procedures in most cases appear to be aimed at gaining legitimacy and consensus more than how much they do not open up spaces for real participation[3]

Criticism on the neoliberal orientation and doubts regarding the democratic character of new forms of urban governance in Global North have also accompanied their spread to the rest of the world, including countries of the Global South, where "the concept of governance has been strongly promoted as a policy measure, along with decentralization, local democratization, driven largely by multi-lateral institutions, such as World Bank and UN agencies" (Watson 2009a: 157).

As reconstructed by Watson, one of the most engaged Global South planning theorists, and also by Smit (2018) in a recent review on urban governance studies in Africa, after the sponsorship in the 1980s of "pure" neoliberal economic policies—focused on privatization, deregulation, and decentralization—it is World Bank for first to launch the concept of *good governance* in the report *World Bank Study Sub-Saharan Africa—from Crisis to Sustainable Growth* (World Bank 1989), a concept taken up and expanded in subsequent reports (World Bank 1992, 1994) until the *World Development Report* 1997: *the State in a Changing World* (World Bank 1997), in which the "importance of strong and effective institutions, rather than the rollback of the state, as in the past" was underlined. Thanks to this evolution: "Since the late 1990s, 'good governance' has become the mantra for development in the South and planning has been supported to the extent that it has promoted this ideal" (Watson 2009a: 158).

---

[3] On the controversial "Janus face" of governance see also Swyngedouw (2005), in References.

A global success once again effectively summarized by Obeng-Odom (2013): "Throughout the world, this is the age of governance, the era of the city and a period of good urban governance".

However, even the pathway of *good governance* in the Global South has encountered differences of interpretation and many obstacles in its implementation. The World Bank approach, which largely focused on efficiency and accountability, has been strongly criticized by other global agencies such as UNDP and UN-Habitat for being "a mainly administrative and managerialist interpretation of good governance" (UN-Habitat 2016: 10).

These agencies have, therefore, promoted a revision of the concept of *good governance* that places the emphasis on democratic practices and human and civil rights, a version spread worldwide through many reports and global campaigns, from the first *UNDP Governance for Sustainable Urban Development* (1997) to the *Global Campaign on Urban Governance* of UN-Habitat in 2002 and so on through many others (UN-Habitat 2002, 2003, 2008, 2009, 2014, 2016), until the maturation of an articulated concept of *inclusive, multi-scale and multi-level governance* in the *New Urban Agenda* adopted at the *United Nations Conference on Housing and Sustainable Urban Development (Habitat III)* in Quito, Ecuador, on 2016 (United Nations 2017a, b).

The UN-Habitat approach, thus, sought to reorient governance strategies to reduce the profound imbalances between political powers and the enormous socio-spatial inequalities of the urban realities of the Global South, supporting processes of decentralization of the urban government and strengthening the management capacities of local authorities, especially with regard to upgrading programs of slums and informal settlements with the participation of local communities.

Along this way, governance strategies have been enriched with consistent capacity building functions in local institutions and local communities, seeking to assume not only a more democratic character, but also a pedagogical purpose.[4]

The influence of the main international agencies on the urban policies of the Global South countries, especially those for housing, as argued by Chiodelli (2016), has been relevant, even if partial and sometimes contradictory: on the one hand, through the financing and direct promotion of intervention programs and technical-training support functions for the definition of policies and projects of central and local governments, these agencies have given impetus to new guidelines and new operational methods of the urban governance. On the other hand, many southern countries have continued to practice policies other than those supported by international organizations (i.e., eviction and demolition of informal settlements), while

---

[4] The relevance assigned to these activities to achieve an effective "good urban governance", is clearly underlined in Habitat III Policy Papers: "Capacity building for urban governance needs to be accelerated: improving differentiated capacities linked to urban governance needs to take into account institutional capacities, the technical and professional skills of individuals as well as local leadership skills. Building capacities related to urban planning, budgeting, public asset management, digital era governance, data gathering and engaging with other stakeholders are of particular urgency. Capacity building actions need to go beyond conventional training and stimulate learning in the short, medium and long-term" (United Nations 2017a, b: 4).

the outcome of the structural intervention programs promoted by the International Monetary Fund and the World Bank in many cases was that, on the contrary, to worsen the housing and living conditions of the slums.

Therefore, even in countries of the South, the spread of governance paradigm in urban strategies, supported by multi-lateral organizations, has been accompanied by many criticisms regarding its effective capacity for innovation of urban government in a democratic way, as well as on the real extent of changes induced by new decision-making methods in expanding urban contexts.[5]

As Watson (2009a, b: 158) finally points out: "In the Global South, as elsewhere, there is a tension between the participative and technocratic dimension of new approaches to governance, as well as between participative and representative democracy", and despite the pressure from international agencies, "actual decentralization, local democratization and shared governance have been uneven processes in the global south and in many parts changes have been limited. Limited capacity, resources and data at the local level have further hindered decentralization" (ibidem). All these critical conditions are still to be found in Africa and above all in the sub-Saharan countries.

2 The Raising of Urban Governance in Sub-Saharan Africa Between Critical and Potentials

The most relevant studies focused on urbanization and urban management in Africa and particularly in Sub-Saharan African countries (Watson 2009a, b; Myers 2011; Jenkins 2013; Pieterse and Simone 2013; Parnell and Pieterse 2014; Obeng-Odom 2017, Smit 2018; Home 2021) highlight, in fact, some "structural" nodes that aggravate all the obstacles encountered in the diffusion of Western urban governance paradigm in the Global South: from state centralism and weakness of local governments to the scarcity of public resources to be allocated to urban and territorial infrastructures; from the overwhelming power of real estate investors and international developers to the predominance of the informal sector in the economy and urban development; from the weakness of organizations of civil society to the ineffectiveness of obsolete planning models inherited from the colonial past; from the inadequacy of technical structures to the scarcity of territorial information, to name only the most cited.

Particularly Jenkins (2013), in the introduction to his in-depth study on the case of Maputo, Mozambique, identifies the main structural problems of sub-Saharan urbanism in the detachment between massive urbanization and economic growth, which is largely concentrated on the extraction and export of natural resources, but limited in general economies:

---

[5] For a critical review on the neo-liberalism paradigm applied to urban strategies in developing countries see in References: Burgess et al. (1997) and also Helmsing (2002) with specific regard to housing policies.

The result is thus likely to be a continuation of the past decades of constraints on wider expansion of what is usually termed the 'formal' economy, in other words that with some form of state regulation and engagement, including taxation. (…) This all means that the precondition for possible urban consolidation through economic growth and some form of wealth redistribution is limited. On the other hand global pressures on agricultural production (especially Western government subsidies) mean that development opportunities are also undermined for many rural dwellers and the outcome is effectively the definitive urbanization of poverty. (Jenkins 2013: 19–20)

Jenkins also highlights very effectively the critical issues of governance in this context:

The limitation on 'formal' economic development is not only due to the nature of global economic interests in the macro-region, but also due to continuing elite nature of the region's governance. This is underpinned by the region's complex political structure of some 50 nation-states in one-fifth of the world's land surface. The relative weakness of these nation-states derives initially from their colonial construction but continues into the fifth or sixth decade of postcolonialism for many countries, due to a range of internal factors (e.g., ethnic competition) as well as external factors (e.g., global economic peripherality). Most governments have relatively weak administrative and technical capacities, and this is particularly the case at local government level, where in many situations local authorities with any form of autonomy and/or democratic political representation are relatively new and still highly depended on the central state. (…) Hence, the capacity to respond to accelerating urbanization from the local government's point of view is extremely limited and highly dependent on central government subsidy and/or foreign investment and international aid. (Jenkins 2013: 19–21)

The weakness of local government is also underlined by Smit (2018) in his overview on the main actors of urban governance in Africa, noting how the impetus given by governments and international agencies toward decentralization from the 1980s onwards has had a very patchy and partial implementation, and in some cases has been overturned. Therefore "it has been argued by some scholar that the rushed and partial decentralization of public authority in Africa has often resulted in local governments that are 'weak, disorganized, inadequately trained and staffed, and often under-resourced relative to the new range of responsibilities they are expected to take on' (Meagher 2011, 51)" (Smit 2018: 6).

Another typical (and critical) feature of African urban governance noted by Smit, concerns the very important role that can play the "traditional leaders", instead of institutional ones, especially as regards the allocation of land in peri-urban areas for the development of informal settlements. Although traditional leaders are controversial figures—due to the "extra-legal" nature of their activities, often marked by corruption—these typical local actors reveal the presence of informal structures of "customary governance"[6] in the management of urbanization processes, which are intertwined with formal governance networks to defend traditional rights and mediate between the two sectors.

---

[6] The concept of "customary governance" refers to the theory of "customary law", the traditional cultural practices that become "laws" parallel to the official ones, originally formulated by Comaroff and Roberts (1981) with references to studies conducted precisely in an African context.

This problematic and often conflicting dualism in the dynamics of urban governance at the local level generally leaves much room for action of supra-local actors: central government, multi-lateral agencies, development banks, international donor agencies, and large sector organizations—such as real estate development or food production companies—and to other emerging actors from the informal sphere, such as informal business organizations usually "governing" marketplaces and streets traders (Brown et al. 2010).

In addition to government and private actors, other scholars (Devas 2001; Olivier de Sardan 2011; Tostensen et al. 2001) highlight the presence of a vast range of civil society associations -ethnicity-based networks, home-town associations, youth associations, savings groups, funeral groups, religious association, etc.- that often, in practice, "perform roles undertaken by the state in cities in the global North, such as providing basic services, allocating land, ensuring safety, providing social security nets, and so on" (Smit, 2018: 8). Many community-based associations have been also set up by international development agencies to implement programs of slum-upgrading, urban and rural agriculture, food security, etc. The fundamental role of civil society associations, and particularly of collaborative fluid networks between informal actors and marginalized residents of African cities in supporting urban functioning is emphasized by authors such as Bayat (2004) and especially Simone (2004). However, several scholars argue that participation of civil society in urban governance, especially in sub-Saharan Africa, remains marginal and a very problematic issue because "(…) social networks which extend beyond kinship and ethnicity remain largely casual, unstructured, and paternalistic" (Bayat 2004: 85) and "participation is still mediated more typically by patron-client relations, rather than popular activism" (Watson 2009a: 159).

This very concise overview of the main and most typical actors and stakeholders involved in urban governance is however sufficient to highlight, on one hand, the inequality of resources and capacities between public, private, and civil society actors, and, on the other, the relevance of the informal governance actors and processes in remedying this disparity. This condition is a common trait to the main sectors of urban development,[7] where the presence of a vast and pervasive informal system, which supports or integrates the more limited formal systems—of the housing market, production and management of urban services, food production and retail in the markets, for example—create a formal/informal continuum often difficult to disentangle, within which, as argues by Devas (2004): "informal governance processes are, in practice, often more important than formal governance process".

Studies on African urbanism, therefore, converge in highlighting some distinctive features of urbanization processes in Sub-Saharan Africa—urban growth mainly disjointed by industrialization, strong urban–rural interconnection, predominance of

---

[7] Smit identifies three key areas of urban governance in Africa: land allocation and land use management, the provision and management of basic infrastructure services, such as water, sanitation, and waste management", and transport/accessibility system, although many other areas are gradually becoming the subject of governance, including environmental and disaster risk management, education and socio-cultural development, etc.

informality in business, housing, and services provision—and the particular limitations of "formal action" of government, at the central and, above all, at the local level, which leave room for uncoordinated and fragmented governance in a vast and often opaque network of formal and informal processes. If these ambiguous governance conditions, on the one hand, "leads to a situation where in fact Sub-Saharan African cities function, albeit in ways which seem chaotic and noncontrolled at first sight", as argued by Jenkins (2013: 7), on the other hand, it inevitably tends to favor the agenda and interests of stakeholders with the most skills and resources—"opening the door" to corruption as well—and to exclude problems and needs that concern the lower sector of society, typically the urban poor and their living environment.

In addition to these more structural problems, many scholars have highlighted the significant role, in supporting such unequal and contested governance framework, played by the Western and "modernist" urban planning system inherited from the colonial past, still in use in many of the Sub-Saharan African countries (Njoh 1999, 2003; Devas 2001; Nunes Silva 2015, 2020).

The negative impact of the inherited planning system "in worsening poverty and the environment" is particularly underlined by Watson identifying the different elements that make the ideas and tools of spatial planning inadequate and ineffective in Africa, because widespread "(...) mainly through British, German, French and Portuguese influence, using their home-grown instruments of master planning, zoning, building regulations and the urban models of the time – garden cities, neighborhood units and Radburn layouts, and later urban modernism", and usually applied only in the central areas inhabited by Europeans (Watson 2009a: 172–178). More recently still used for the new towns and "green enclaves" for richer classes and the new emerging middle class (Mazzolini 2016a, b), the modernist and "rational-comprehensive planning" tradition continued to reinforce spatial and social exclusion, as well as the unsustainable urban sprawl.

The recognition of the substantial "diversity" of urban transition in Sub-Saharan Africa, including the peculiar tangle between formal and informal decision-making processes that "govern" such transition, has prompted a growing number of scholars (between many others: Roy 2016; Watson 2014, 2016; de Satgé and Watson 2018; Bolay 2020) to criticize the "idealistic" and largely inapplicable approaches to planning of the past, as well as the uncritical adoption of the "new wave of context-less planning ideas" designed for northern cities in more recent times, because "This new era of planning (using terms such as eco-cities, smart cities and world-class cities) is again imposing a concept of 'good cities' derived from other and very different contexts" (de Satgé and Watson 2018).

To overcome these old and new Western legacies, a "Southern Urbanism" theory is raising, specially tailored on distinctive characteristics of African urbanism, to be considered as opportunities for innovative and more effective forms of planning and urban governance strategies.

3   Toward New Paradigms and Urban Research Approaches for Alternative Urban Governance Framework in Sub-Saharan Africa: The Case of the "Boa_Ma_Nhã, Maputo!" Action-Research

The need to decolonize the urban planning system in Africa, and to develop new approaches, methods, and institutional capacities for urban governance, calibrated to the specificities of the African city in its various regional variations, is therefore long recognized and claimed by a growing number of international and African theorists (in particular: Myers 2011; Pieterse and Simone 2013; Parnell and Olfield 2014; Mabin 2014; Parnell and Pieterse 2015; Hyman and Pieterse 2017; Simone and Pieterse 2017).

However, the process of decolonization and "locationally innovation" of planning and urban governance, argue many of these scholars, require, first of all, a specific knowledge production on African urbanism, rooted in a new theory and praxis of urban research in that context. In other words, it is necessary, as suggested by the title of Parnell and Pieterse (2015), a deep "rethinking of methods and modes of African urban research", that is a new research approach, named by these authors "translational global praxis", which "captures more than the idea of applied research or even co-production, and encompasses integrating the research conception, design, execution, application and reflection - and conceiving of this set of activities as a singular research/practice processes that is by its nature deeply political and loca-tionally embedded". To understand the specificity of African urbanism and therefore "For knowing what can be done to affect the change of the city", it is imperative to adapt the methods and modalities of African urban research to the conditions of the context, "where human needs are great, information is poor, governance conditions are complex, and reality is changing".[8]

More specifically, the "fragile" conditions of local institutions, civil society, and academia that characterize many African countries make it necessary to adopt a strongly engaged action-research approach, that involves all these components in a "translational" process connecting research, policy, and practice to support urban change.

This appeal to the commitment of urban research in the African urban context has been increasingly welcomed by the international scientific community, thanks also to the growing academic cooperation that often integrates cooperation for development, promoting partnerships with universities, institutions and other stakeholders of the African countries (Petrillo and Bellaviti 2018a, b).

Through the blending of the three pillars of academic activity: research, training and know-sharing, universities can, in fact, contribute to develop trans-disciplinary action-research, supported by consistent multi-level capacity building initiatives in

---

[8] Parnelle and Pieterse's reflection is based on the experience of the African Center for Cities (ACC) at the University of Cape Town (www.africancentreforcities.net), the main research and training hub on Southern urbanism theory and praxis.

academia and local institutions, involving also economic and social realities in these processes of co-production of knowledge and action capacity.

An example of this great potential is represented by the "Boa_Ma_Nhã, Maputo!" action-research reported in this book, carried out through the extensive collaboration between Politecnico di Milano with the Mondlane University of Maputo, the Italian Agency for Development Cooperation and many other local institutions and NGOs.

Conceived and designed to offer a support of knowledge and guidelines for the promotion of a governance framework of the "unknown Metropolis" that have been shaping in the past decades in the outskirts of the capital of Mozambique, fragmented in terms of administrative boundaries and governance and shaped by a complex tangle of informal or unmapped flows and systems, the study tackles many of the obstacles mentioned above with an approach that "breaks" with the traditional forms of planning still prevailing, to tune in with new conditions, criticalities, and opportunities present in the territory of investigation.

For such a demanding research project, the first problem to be addressed was the lack of information regarding existing cross-scalar patterns that have been shaping this territory, the scarcity and inconsistency of the available statistical data, the lack of cartographies, and the lack of investigations of economic related transformations.

The construction of a broad and updated territorial framework, through a transdisciplinary research program, which integrates disciplines such as architecture, urban planning, hydraulic, energy, and computer engineering, and combines quantitative analysis with qualitative research and participatory methodologies (infield investigation, interviews, and focus groups with local stakeholders, case-studies), thus represents an essential element to start the construction of a new sustainable development scenario for the area, at the same time intercepting the potential stakeholders of a new territorial "soft governance", released from the limits imposed by the elephantine, compartmentalized and rigid administrative structures of colonial legacy.

The research activities, focused in particular on the Water-Energy-Food Nexus, considering the potential evolution of the agriculture sector, backbone economy of the area, and the whole food cycle and its multiple environmental, economic, social, cultural implications, are at the same time an instrument of knowledge and representation of the territory, and a device for identifying and activating the new structures of territorial governance, built *ad hoc* in relation to the local development plans, which are entrusted with the implementation of the scenario hypotheses. It is inside these plans, and related pilot projects, that the very local specificities of territories and of stakeholders and partners find their place, including the customary informal structures of governance already present in the territory, in search of alternative "hybrid governance" frameworks to guide and facilitate metropolitan growth, through the management of natural resources and large-scale infrastructure as collective assets.

On this innovative platform of updated and trans-disciplinary knowledge, built with the contribution of local stakeholders, a wide action of capacity building at multi-level can be developed:

- Several knowledge-sharing initiatives are developed as academic cooperation initiatives: teaching activities at the FAPF-UEM newly established Master Level Course and Ph.D. program; exchange activities of the researchers; dissemination activities, such as workshops and training events organized at the local level, to reach different audiences.
- New tools for territorial knowledge are made available to local institutions— Assessment analysis and Scenario models—and specific guidelines to support decision-makers dealing with the challenges of sustainable development in fragile contexts of the Global South, such as Mozambique.
- Finally, with the Pilot project method, capacity building action spreads to the widest range of stakeholders involved in the first initiatives for the implementation of development plans, investing in education, and local rural entrepreneurship with the aim of producing measurable impacts.

There is no need here to further investigate the "Boa_Ma_Nhã, Maputo!" action research—amply documented and argued in other essays in this volume—to underline, in conclusion, how university cooperation initiatives such as this one can give a great impetus to the innovation of territorial governance and urban management in contexts such as African countries, unhinging the obsolete systems of the colonial legacy and searching "on the field" more adaptive formulas to local and specific systems.

The Science Diplomacy that universities carry out, with their own prerogatives— research, training, know-sharing, capacity building—can indeed operate across the different sectors and levels of governance—formal and informal, central and very local—and build new networks that are more inclusive and locally rooted, in search of solutions consistent with the ideals of sustainable urban transition (inclusive resource-efficient, affordable and low-carbon) but at the same time compatible with practices and knowledge expressed locally, closer to real possibilities and potential of the territories and cities of Africa—and more generally of the Global South.

References

Bayat A (2004) Globalization and the politics of the informals in the Global South. In: Roy A, Alsayyad N (eds) Urban informality: transnational perspective from the Middle East, Latin America and South Asia. Lexington Books, Oxford

Bolay JC (2020) Urban planning against poverty. How to think and do better cities in the global South. Springer Open, Cham

Brenner N (1999) Globalisation as reterritorialisation: the rescaling of urban governance in the European Union. Urban Studies 36(3):431–451

Brenner N (2004) New state spaces. Urban governance and the rescaling of state-hood. Oxford University Press, New York

Brenner N, Theodore N (eds) (2002) Spaces of neoliberalism: urban restructuring in North America and Western Europe. Blackwell, Oxford

Brown A, Lyons M, Dankoco I (2010) Street traders and the emerging spaces for urban voice and citizenship in African cities. Urban Stud 47(3):666–683

Burgess R, Carmona M, Kolstee T (1997) The challenge of sustainable cities. Neoliberalism and urban strategies in developing countries. Zed Books, London-New Jersey

Comaroff JL, Roberts S (1981) Rules and processes: the cultural logic of dispute in an African context. University of Chicago Press, Chicago

Chiodelli F (2016) International housing policy for the urban poor and the informal city in the global South: a non-diachronic review. J Int Dev 28(5):1–20

de Sardan JPO (2011) The eight modes of local governance in West Africa. IDS Bull 42(2):22–31

de Satgé R, Watson V (2018) Urban planning in the global South: conflicting rationalities in contested urban space. Palgrave Macmillian, London

Devas N (2001) Does city governance matter for the urban poor? Int Plan Stud 6(4):393–408

Devas N (2004) Urban governance, voice and poverty in the developing world. Earthscan, London

Helmsing AHJ (2002) Decentralisation, enablement, and local governance in low-income countries. Environ Plann C Gov Policy 20(3):317–340

Home R (ed) (2021) Land issues for urban governance in Sub-Saharan Africa. Springer, Cham

Hyman K, Pieterse E (2017) Infrastructure deficits and potential in African Cities. In: Burdett R, Hall S (eds) The Sage handbook of urban sociology: new approaches to the twenty-first century city. Sage Publishers, London

Jenkins P (2013) Urbanization, urbanism and urbanity in an African City: home, spaces and house cultures. Palgrave Macmillan, London-New York

Mabin A (2014) Grounding Southern city theory in time and place. In: Parnell S, Oldfield S (eds) (2016) Routledge handbook on cities of the Global South. Routledge, London

Meagher K (2011) Informal economies and urban governance in Nigeria: popular empowerment or political exclusion? African Studies Review 54(2):47–72

Myers G (2011) African cities: alternative visions of urban theory and practice. Zed Books, London and New York

Mazzolini A (2016a) An urban middle class and the vacillation of 'informal' boundaries-insights from Maputo, Mozambique. Geograp Res Forum 36:68–85

Mazzolini A (2016b) The rising "Floating class" in Sub-Saharan Africa and its impact on local governance: insight from Mozambique. In: Nunes Silva C (ed) Governing urban Africa. Palgrave Macmillian, Basingstoke, pp 213–246

Njoh AJ (1999) Urban planning, housing and spatial structure in Sub-Saharan Africa. Ashgate, Aldershot, UK

Njoh A J (2003) Planning in contemporary Africa: the State, town planning and society in Cameroon. Ashgate, Aldershot, UK

Nunes Silva C (2015) Urban planning in Sub-Saharan Africa. Routledge, London

Nunes Silva C (ed) (2020) The routledge handbook of urban planning in Africa. Routledge, London

Obeng-Odom F (2013) Governance for pro-poor urban development. Lessons from Ghana. Routledge, London

Obeng-Odom F (2017) Urban governance in Africa today: reframing, experiences, and lessons. Growth Change 48(1):4–21

Olivier de Sardan J-P (2011) The eight modes of local governments in West Africa. IDS Bull 42(2):22–31

Parnell S, Oldfield S (eds) (2016) Routledge handbook on cities of the global South. Routledge, London

Parnell S, Pieterse E (eds) (2014) Africa's urban revolution. Zed Books, London

Parnell S, Pieterse E (2015) Translational global praxis: rethinking methods and modes of African urban research. Int J Urban Reg Res 40(1):236–246

Petrillo A (2017) La governance dell'informale tra l'eccezione e la norma. In: Paone S, Petrillo A, Chiodelli F (a cura di) Governare l'ingovernabile. ETS, Pisa

Petrillo A, Bellaviti P (2018a) Sharing knowledge for change. Universities and new cultures of cooperation: transnational research and higher education for sustainable global urban development. In: Petrillo A, Bellaviti P (eds) Sustainable urban development and globalisation. New strategies for new challenges-with a focus on the global South. Springer, Cham

Petrillo A, Bellaviti P (eds) (2018b) Sustainable urban development and globalisation. New strategies for new challenges-with a focus on the global South. Springer, Cham

Pieterse E, Simone A (eds) (2013) Rogue urbanism: emergent African cities. Jacana Media and African Center for Cities, Johannesburg

Roy A (2016) Who's afraid of postcolonial theory? Int J Urban Reg Res 40(1):200–209

Simone A (2004) For the city yet to come: changing African life in four cities. Duke University Press, Durham, N.C.; London

Simone A, Pieterse E (2017) New urban worlds: inhabiting dissonant times. Polity Press, Cambridge, UK

Swyngedouw E (2005) Governance innovation and the citizen: the Janus Face of governance-beyond-the-state. Urban Stud 42(11):1991–2006

Smit W (2018) Urban governance in Africa: an overview. Int Develop Policy—Revue international de politique de développement 10:55–77

Tostensen A, Tvedten I and Vaa M (eds) (2001) Associational life in african cities: popular responses to the urban Crisis. Nordiska Afrikainstituted, Uppsala

UNDP (1997) Governance for sustainable human development. UNDP, New York

UN-Habitat (2002) Global campaign on urban governance: concept paper, 2nd edn. UN-Habitat, Nairobi

UN-Habitat (2003) The challenge of slums. Global Report on Human Settlements 2003. UN-Habitat, Nairobi

UN-Habitat (2008) The State of African Cities 2008: a framework for addressing urban challenges in Africa. UN-Habitat, Nairobi

UN-Habitat (2009) Global report on human settlement 2009: planning sustainable cities. Earthscan, London

UN-Habitat (2014) The state of African Cities 2014: re-imagining sustainable urban transition. Earthscan, London

UN-Habitat (2016) Urbanization and development: emerging futures. World Cities Report 2016. Un-Habitat, Nairobi

United Nations (2017a) New urban agenda. www.habitat3.org. Accessed 11 Dec 2021

United Nations (2017b) Habitat III policy papers: policy paper 4 urban governance, capacity and institutional development. United Nations, New York. www.habitat3.org. Accessed 11 Dec 2021

Watson V (2009a) 'The planned city sweeps the poor away...': urban planning and 21st century urbanization. Prog Plan 72:151–193

Watson V (2009b) Seeing from the South: refocusing urban planning on the globe's central urban issues. Urban Stud 46(11):2259–2275

Watson V (2014) African urban fantasies: dreams or nightmares? Environment and urbanization 26(1):215–231

Watson V (2016) Shifting approaches to planning theory: global North and South. Urban Plan 1(4):32–41

World Bank (1989) Sub-Saharan Africa, from crisis to sustainable growth, a long term perspective study. The World Bank, Washington, DC

World Bank (1992) Governance and development. The World Bank, Washington, DC

World Bank (1994) Governance, The World Bank's experience. The World Bank, Washington, DC

World Bank (1997) World development report 1997: the State in a changing world. The World Bank, Washington, DC

Mobility: Developing Countries Through the Lens of Megaprojects, Equity, Sustainability, and Development

Paolo Beria

Abstract Mobility and transport investment in developing countries face different challenges from those of developed economies. A viable and balanced recipe for them should skip a Euro-centric approach and focus on local conditions. In particular, the role of megaproject must be questioned, as they are often risky and out-of-scale, fitted for extraction and industrial sectors that typically are aliens to local economy and development. In this chapter I propose a reflection on lighter solutions for mobility problems in Africa, focusing in particular on habilitating conditions: a reliable and diffused road and air accessibility, maintenance, the support to investments in vehicles, feasible solutions to traffic in cities. In the second part, I focus on two very different cases of transport megaprojects in Mozambique: the Maputo Corridor and the Maputo-Catembe bridge; pointing out that the first is a successful case because solving a real missing link and part of an industrial strategy, while the second is a perfect *white elephant*.

1 Introduction

Global mobility challenges are well known: climate change and sustainability, congestion, energy consumption, excessive car ownership and land use related problems, economic impacts of mobility and accessibility, liberalization of markets, etc. But moving the discussion to the mobility challenges in developing countries requires also considering the specific needs and problems of such countries, as well as the actual contribution they can fairly give to the achievement of global goals. Climate change and pollution are a threat for developing countries, but their contribution (in the transport field) to the solution of the global problem is, and likely must remain, limited. Congestion is also an issue, but for reasons that are totally different than in developed countries. Excessive car ownership can be a problem in some large cities, but overall, the availability of vehicles is scant and this constitutes a dramatic

P. Beria (✉)
Department of Architecture and Urban Studies, Politecnico Di Milano, Milan, Italy
e-mail: paolo.beria@polimi.it

© The Author(s), under exclusive license to Springer Nature Switzerland AG 2022
L. Montedoro et al. (eds.), *Territorial Development and Water-Energy-Food Nexus in the Global South*, Research for Development,
https://doi.org/10.1007/978-3-030-96538-9_16

223

problem for economic development rather than a goal like in western or Asian countries. Inaccessibility, in general, can be a severe limit to the economic development of territories, populations and specific categories (women, for example).

In the first section of this chapter, I discuss some of the mobility challenges referring to developing countries, trying not to adopt a Euro-centric vision of transport and mobility. In the second part, I focus on Mozambique and the Maputo Province, in particular, presenting and comparing two recent megaprojects (the Maputo Corridor and the Maputo-Catembe bridge) and discussing possible alternative transport policies for the country.

2 Mobility Challenges in Developing Countries

2.1 (Large) Projects as a Promise of Development

The keyword for transport planners in developing countries is often "investment". The insufficiency or the obsolescence of infrastructure stocks is often dramatic, and *inaccessibility* is a word that has a totally different sense than in Europe or North America, where the focus is on travel time and cost reduction, but surely not on the practical *impossibility* to reach a place and settle economic activities in there.

The logic of infrastructure investments, everywhere, is twofold, but the two layers are often overlapping, confused, and misunderstood: a large injection of money into an economy, whatever is the goal of the investment, and the means to reduce transport costs, whose second-order consequences should be an advantage for firms and citizens. Politicians tend to focus on the first, which is an excellent trophy to exhibit for the short term. Planners should focus on the second. Economists should focus on both, even if many tend to consider the second as ancillary.

In developed countries, the injection of money comes from the public, often in form of debt. It is a problem, but of different nature. In developing ones, where access to debt is limited, an injection could come from outside, in the form of foreign investments. But the logic of foreign investment is different from local public ones: profit (which is good, if obtained in a healthy economic environment) for the first, fulfillment of needs (hopefully) for the second. The challenge is that the two match, but it is far from guaranteed, especially if the contract is asymmetric: a large foreign investor promising money in change of land or resources (and often also subsidies), and a local government without comparable financial alternatives that hopes in future economic spillovers from it.

Concerning the direct effects of a transport investment, they hugely vary according to the investment itself. Typically, large foreign investments go to infrastructure to support the primary and industrial sectors (ports for oil and gas, but also railways for mines and large industries). The wish of local governments is that these investments (the transport ones as a component of the industrial ones) will also boost the local economy, including the creation of jobs. It is what happened during the industrial revolution, when transport investment was part of industrial development and accompanied the growth of western economies. But nowadays economy is different and these spillovers are not guaranteed or could be extremely limited with respect to the overall value. Moreover, in developing countries the capability of megaprojects to bring local benefits is limited by the lack of productive factors, technology, and workforce that must be imported (Castel-Branco 2008; Mussagy 2014).

So, the challenge should be to design and promote transport infrastructures that do not serve only the global industry, but also the local economy and the people. Unfortunately, this is wishful thinking if looking at major transport megaprojects in Africa, that look like aliens landing on a planet with totally different needs, or "isolated islands" according to Castel-Branco (2008). Billions that could revamp the entire network of a country are spent just to connect one mine.

Megaprojects entail an additional risk: to become political projects, desirable beyond any rational criterion. A *white elephant* is a metaphor for an oversized project, extremely expensive, that gives limited benefits with respect to the effort paid (Beria et al. 2018). *White elephants* exist because politics likes them and creates a narrative on the incredible benefits that will occur in the future, making them political projects and forgetting to manage the real needs. Again, the recent history of Europe is full of *white elephants* that should not be replicated in countries that cannot afford to waste even a cent.

2.2   Equity and Opportunities for All

Today we know that transport infrastructure and mobility services are habilitating elements, necessary conditions for an economy to exist. Megaprojects could contribute to this, but normally their costs and benefits suggest that—whenever available to a developing country—resources could also be spent in more diffused actions. This often means roads, buses, possibly airplanes, but hardly rail lines whose role is shrinking also in developed countries.

A balanced transport policy, not focusing on visionary megaprojects, but aiming at an equitable and decent level of accessibility for people and enterprises, is the challenge for Africa. We can list five ingredients, which do not stimulate any imaginary, but may ground economic development.

(a)   *Roads.* A road network is all that is needed in most territories and economies, including developed countries. Roads cost definitely less than any other infrastructure. Roads make land productive, letting production factors to enter and products to flow out. Roads are used by private and public vehicles. Road circulation is mostly self-regulating.

(b)   *Maintenance and resilience of the network.* Infrastructure decay over time and in some environments decay even faster. A road network not maintained and degraded is not working. It is increasing transport private costs,[1] damaging vehicles, causing accidents. It is not guaranteeing the accessibility it was built for.

(c) *Trucks and buses.* Vehicle availability might be a problem, especially for local productive activities. For basic productions and trade, based on individual work and limited technology, transport could represent the most severe limitation because requiring a capital that could be simply not available. So, a transport policy in poorer country areas could focus on supporting vehicle ownership (loans, subsidies, cooperatives) and this could really become a habilitating element.

(d) *Airports and air services.* Long-distance mobility is necessary to support economic relations, students mobility, migrants. In absence of rail services, buses are clearly the typical solution, which means extremely long travels. This could be a limitation for trade, agglomeration economies, and for middle-class mobility, especially in countries with very long distances and low density. Air transport proved to be an extremely efficient and cheap system also in developing countries, if conditions exist, in particular a truly liberalized environment letting companies to compete and a sufficient capacity at airports. Southeast Asia is probably the most interesting case of air transport success out of Europe.[2]

(e) *Urban transport.* The urban population is increasing and the pressure on cities is exponential, in terms of congestion, pollution, safety, urban quality, and quality of life. Therefore, the mobility needs of African cities are like Asian or European ones in terms of public transport, congestion regulation, and pollution control. However, many differences also exist. The low investment capacity of cities limits the space for rail mass transit and suggests other forms such as Bus Rapid Transit (BRT). The huge number of taxi-like services (from rickshaws to motorbikes, to vans) means that any modal shift will severely affect the low-skill employment that such systems guarantee. The impossibility to reduce travel times to a reasonable level by means of transport measures (because of city size, low car ownership, and no rail) suggests that only a more balanced land use mix might relieve the most disadvantaged from wasting hours in traffic every day to reach the workplace.

---

[1] Vehicles suitable for off-road cost more, require more maintenance, are less fuel-efficient.

[2] ASEAN Single Aviation Market agreement (ASAM) consists of air freedoms up to the 5th for all ASEAN countries and carriers. This level has been reached in 2015 and foresees no fares regulation and no constraints on entry airports. An agreement including a partial 5th freedom is in place also with China (uncongested airports only) and led to significant market entries. Overall, both the internal and Chinese ASEAN markets grew significantly, through a rise of low-cost models (Laplace et al. 2019; Lee 2019). Correlation with GDP growth is found (Laplace and Latgé-Roucolle 2016), but causation is not demonstrated due to the size of GDP and its exogenous growth in the same period (Hakim and Merkert 2016). In many cases, this corresponded to the (relative) decline of the former flag carriers (e.g. Vietnam Airlines, Garuda Indonesia), which lost most of their market dominance (O'Connor et al. 2020).

2.3   Safety, Security, and Sustainability

Road transport for all is the most economically feasible solution for developing countries, but this is apparently not going in the direction of sustainability. However, some arguments must be considered.

a.   The contribution of developing countries to energy consumption is very limited. Any renounce for them costs in terms of underdevelopment but will have a negligible global impact. If the average African citizen would *double* his/her energy consumption to develop, this increase could be offset by a reduction of 13% of a European or just 5% of a North American citizen (see Fig. 1). Everybody should contribute to energy saving, but the poorest should be the last in the list as they have fewer alternatives, but especially higher marginal gains from development.

b.   Technological improvements can potentially be applied everywhere. For example, the cheapest car an African can buy today is surely less polluting than the cheapest car a European had bought 10 or 20 years ago.

c.   Traffic pollution is a problem mostly in large cities, as the low density of emittants and receptors typical of sparse areas makes the negative impacts negligible. So, the use of older cars and trucks is not a major pollution problem outside cities.

d.   Accidents are clearly a problem, and a high toll is paid every year. However, the solution is not a modal shift, but rather an appropriate maintenance of the roads, which is expensive, but gives also economic effects (see previous arguments). Also, modern vehicles, protections, and road education largely contribute.

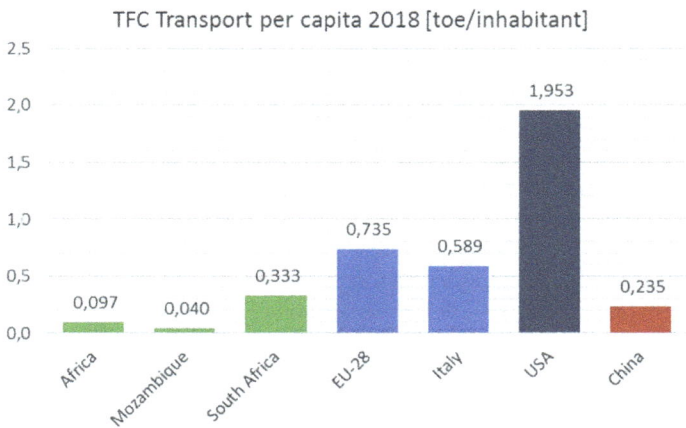

Fig. 1 Total final consumption per capita, in Tonnes Oil Equivalent (toe). *Source* our elaborations on IEA World Energy Balances 2020 and The World Bank, World Development Indicators

e. Reliability of road network is a major issue in tropical areas with unpaved roads. Generally, they increase accidents and damage of vehicles, but sometimes also make inaccessible areas for months, preventing any economic activity.

3 Mozambique and the Maputo Province

The recent investments in the province of Maputo provide an excellent case of a transport policy concentrated on foreign investment megaprojects, instead of one dedicated to backbone the local economy. However, the two cases presented in the following sections are similar for cost and ambition but radically differ in their characteristics and outcomes. The section and the essay will be concluded by a comment on a different policy attitude, more focused on local economy needs.

3.1 The Maputo Corridor: The Opportunities of Solving a Missing Link

The *Maputo Corridor* refers to a complex of infrastructures, including the EN4/N4 toll international motorway, the Ressano Garcia rail line, the related border post, the redevelopment of Maputo and Matola ports, and an 860 km gas pipeline (Ross et al. 2014). Together they connect the South African regions of Gauteng, Mpumalanga, and Limpopo, eSwatini, with southern Mozambique and Maputo in particular. This infrastructure does not only represent a transport facility, but the platform for the industrialization of southern Mozambique, involving also the Mozal aluminum factory in Maputo/Matola, and the gas extraction in Pande and Temane gas fields, plus other productive areas. The corridor is the outcome of a public–private partnership, the *Iniciativa Logística do Corredor de Maputo* (MCLI), initiated in 1994 to give institutional support to the corridor development.

Today, most of the infrastructure is complete and operational, according to different procurement schemes. The toll motorway is a Build-Operate-Transfer concession between the two governments and a private company, the Trans African Concessions (TRAC), lasting for 30 years from 2002. Overall, the concession refers to 630 km, 50 of which consist of a new motorway in Mozambique and the remaining in the rehabilitation of the existing infrastructure in South Africa, from the border to Pretoria. The rail line was initially awarded for a 15 years concession to a consortium between NLPI/Spoornet (51%) and CFM (49%). However, in 2006 after some years of delay, the mandate for rehabilitation of the Mozambican section was given entirely to the national railways CFM. The South African partner, in the meantime renamed Transnet Freight Rail, remained in charge of providing rolling stock (Sequeira et al. 2014). The port of Maputo is a concession, too. It originally lasted for 15 years since 2000 and was granted to a mixed company, 49% owned by Mozambican rail CFM and the remaining 51% by a private consortium. In 2010 the concession period was

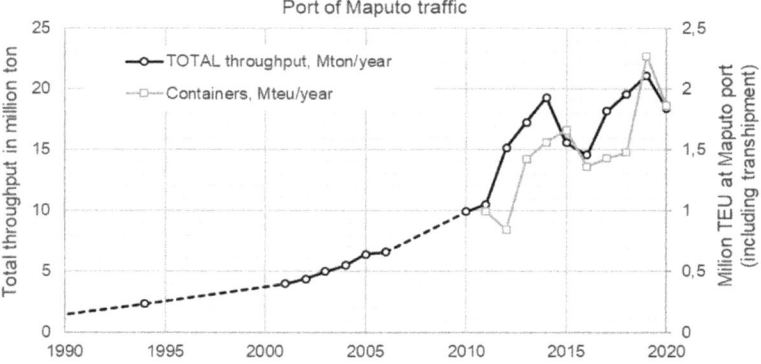

Fig. 2 Port of Maputo traffic. *Source* CFM (until 2006), MPCD Port of Maputo (since 2010)

extended for another 15 years, with an option of an additional ten years of operations after 2033. Today the concessionaire is Maputo Port Development Company (MPDC) and CFM still retains 49% of its shares. Finally, the Mozambique-South Africa Pipeline is operated by a Mozambican company (ROMPCO), a Mozambican-South African joint venture.

Despite the evident importance of the corridor to connect to the sea the landlocked industrial areas of South Africa, it was basically unused until the 2000s due to the interruption of the civil war and consequent disruption of all previous facilities. The interruption forced the flourishing South African industrial sector to develop and use the national port of Durban, which is however 1.5 times farther. After the progressive reopening of the infrastructures, traffic began to grow in all components of the corridor.

The Port of Maputo saw a considerable increase in activities, both as a transshipment terminal and as a hinterland port. Hinterland traffic is fed both by the industries in the metropolitan area of Maputo, MOZAL primarily, and by the international traffic from road and rail connections. The port saw (Fig. 2) a step increase since 2011, a few years after the rail line was reopened, and passed from 10 to 20 Mton/year by 2014. After a decrease, it restarted to grow until 2019 with 22 Mton, until the fall due to the global COVID crisis.

Road traffic on the motorway started to increase steadily since 2006. In 2014 daily vehicles at Maputo toll gate were nearly 60,000, of which nearly 5000 were trucks. Interestingly, also the second toll gate of Moamba saw an increase in traffic, but later since 2009. In 2014 reached the level of nearly 2000 trucks/day (Sequeira et al. 2014). Unfortunately, no newer figures are available. Despite good traffic figures, the financial sustainability of the PPP was at threat because of higher maintenance costs also due to the practice of overloading trucks.

Rail traffic is known in more detail. In Fig. 3 we update the figures present in Sequeira et al. (2014), obtaining a quite different picture in the last years. In 2017

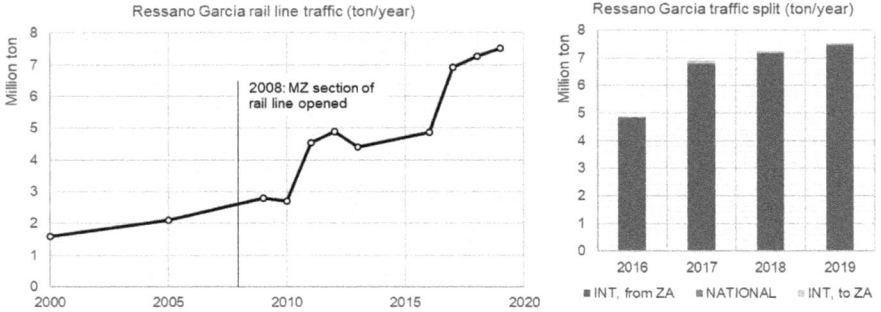

Fig. 3 Ressano Garcia rail line traffic (ton/year). *Source* Sequeira et al. (2014) until 2013, CFM (2017, 2019) since 2016

traffic increased significantly, passing from 5 to 7 Mton per year,[3] with a step comparable to that observed in 2011–2012 when passed from 3 to 5 Mton. It is worth noticing that national traffic is down to zero and traffic from Mozambique (port included) and South Africa counts for just 0.5% of total and decreasing. So, the traffic along the corridor is almost entirely related to South African exports, both to Mozambique and to the port.

The transported goods mix reflects the exports of South Africa and the characteristics of freight rail transport (low-value freight), but also the dependency of Mozambique from food and general goods imports (Fig. 4). Despite growing traffic, rail still represents a negligible share of total trade between the two countries (less than 1% in weight). The reasons are known. In general, rail is intrinsically less attractive than (overloaded) trucks for most goods. It results in a viable option for operators only for intermodal transport and to connect larger industrial facilities, but, in all cases, the quality of the infrastructure and organization must be excellent to offset the limited speed and capillarity of rail. In addition, further local elements of weakness for this line have been pointed out: regulatory problems (absence of competitors to the incumbent companies, no transparency on pricing), technical problems (lack of rolling stock, still inadequate connections in the port), and even political (persistent support to Durban corridor and port) (Sequeira et al. 2014). The fact that the largest amount of goods transported on the line is minerals, and not manufactured goods like metals, goes in this direction.

Transport megaprojects, and megaproject in general, have shown many limits, both in developed and developing economies. The situations where a megaproject became an undeniable turning point or an essential ingredient in the development of an area are scant and often drawbacks prevail, as shortly recalled in the introduction. However, the Maputo Corridor presents two key ingredients of success *as a transport infrastructure*, that are barely present in other comparable cases:

a.   The corridor is not only a transport project but involves also industrial initiatives. This guaranteed financial and political support, but also the transport demand

---

[3] The goal of Transnet Freight Rail is to reach 21 Mton on the corridor (TRANSET 2019).

Volume of ZA to MZ traffic on the RG line, by product category

Value of ZA to MZ traffic on the RG line, by product category

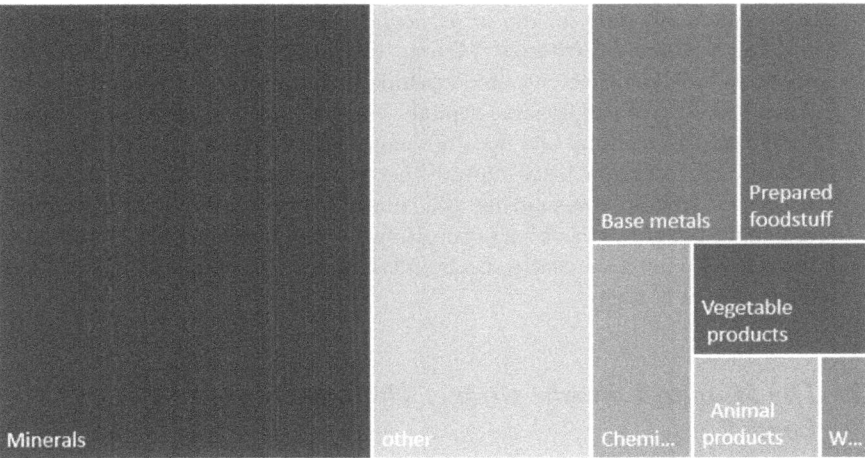

Fig. 4 Ressano Garcia rail line product mix in weight and value in 2020. *Source* our elaborations from CFM (2019)

for the first years and certain coordination of actions. For example, the land infrastructure, which is the most expensive component, came together with the reorganization of the Matola and Maputo ports and shortly after the Mozal smelter.

As suggested by Sequeira et al. (2014, p. 21):

This initiative represented a break with the past as its stated goal went beyond just building isolated infrastructure, but to cater to all transport and logistics needs of a vast cross-border region with high-growth potential in industry, trade, agriculture, tourism and mining.

b. The corridor, in its Mozambican part, represents the removal of a *real* bottleneck: on the one side the landlocked and largest industrial concentration of southern Africa, plus some mining and forestry spots, on the other an existing port extremely well provided in terms of natural protection and canal depth. Moreover, this corridor and port are, or should be, alternative to a longer path through the port of Durban, allowing (at least potentially) significant cost savings.

Aside from these elements—that represent important necessary conditions for success—some issues remain unsolved.

a. The two land infrastructures are in competition, and road transport is often preferred by transport companies for its flexibility and reliability. Rail transport, moreover, suffers from the lack of intramodal competition, with one national incumbent suspected to prefer other corridors (TFR) and the other not particularly proactive in attracting traffic (CFM).
b. The freight traffic is mostly one-way, from South Africa to Maputo and port, which means that capacity and costs are not optimized. In general, it proved not to be useful to support the inland exports of Mozambique but apparently also the imports of South Africa.
c. The corridor, and the rail line in particular, gave limited spillovers into the economy of crossed territories. Mining, extraction, and the large industrial complexes in Maputo belong to international investors and drain most of the qualified workforce and national capitals. But the local Mozambican economy is essentially agricultural and does not gain much from the rail and the international port, moreover if its transport needs are neglected (Stein and Kalina 2019). The corridor works mostly as a "tunnel" from South Africa to Maputo, crossing the other provinces but not creating any substantial relation. Moreover, it proved to be not particularly successful in promoting the settlement of local industry out of Maputo.

3.2 The Maputo-Catembe Bridge: The Opening of "New Lands"

Of totally different nature is the one of the Maputo-Catembe bridge, the second megaproject of the area of Maputo, the longest bridge of entire Africa, and the most expensive project in Mozambique since independence (750 M$). The bridge has been built thanks to Chinese *loans* and increased the financial exposition of Mozambique towards China.

Despite its size and cost, the bridge connects the capital with the District of Matutuíne, an empty region with just 43,000 inhabitants (2017), no industrial production, and no border posts or even roads with eSwatini and South Africa. The district is basically an isolated greenfield. Consequently, there is no transport demand, neither potential, to justify more than simple roads. In addition, the bridge is not a missing link of a network, since there is no network, neither road or rail, to be mended and a possible freight corridor to South Africa, to be entirely built, would be useless since the more than 500 km to Durban can be more economically ran by sea.

So, why has the bridge been built? From the investors' point of view, the operation should probably be risk-free, assuming that they obtained sufficient guarantees from the Mozambican government as always happened in similar cases. One ending is that it is a perfect example of a *white elephant*: gigantic, useless, and very political. However, the expectations of the parties involved are that the Bridge is a necessary means to extend the Maputo urban area into the "greenfield" of the bay's south shore. The bridge in fact is a precondition for the KaTembe residential area project led by the BETAR Group: more than 4000 ha of urban development, including 54% of social housing (Coutinho 2018). The development is included in the General Urbanization Plan of the KaTembe Municipal District (PGUDMK), whose zoning foresees 400,000 inhabitants, schools and health facilities, industrial zones, green areas, and parks (Herzog 2019). Three years after the Bridge opening, the lack of actual implementation mechanisms has generated a mixed situation, with some demarcation of zones started in farther areas, but also the fill and extension of informal settlements in open lands.

Apart from the projects for Catembe, the Bridge may become the stimulus, wanted or not, for more extensive land grabbing activity in the entire territory of Matutuíne, characterized by a particularly sensitive environment.

3.3 Air Transport for All

A more viable solution for Mozambican long-distance mobility is hardly the construction of very long land infrastructures—that in any case would mostly run from the West to the sea and serve the primary sector rather than population and local economy—but the development of a broader air transport network. To date, the Mozambican air market is very limited, prices high and connections infrequent. Aero-taxi is widespread but serves the mining sites or isolated tourism spots and is totally unsuitable for middle-class and local business traffic.

The ingredients for a commercial air transport development are four:

1. Long distances and population concentrated in cities.
2. Capacity at airports capable to host mid-sized narrow body aircrafts.
3. Capitals for the aircrafts.
4. A liberalized environment, open to cabotage from international airlines.

The spatial structure seems favorable: despite the large rural population, Mozambican cities are relatively large and compact, with 14 centers with more than 100 k inhabitants already in 2007 and growing. Many of them already have, or are reasonably distant from, an airport with an asphalt runway. Instead, almost all minor and isolated towns have an airfield, which is however not suitable for commercial aircrafts. The size of the air market is, today, very limited and concentrated on business trips.[4] However, air transport does not need very large numbers and hundreds of travelers per day can justify a route, and the limitedness of land transport together with distances significantly reduce the intermodal competition.

The remaining two conditions are instead more problematic. The national company LAM is basically a monopoly and the recent entry of Ethiopian Airlines into the domestic market was promising but has been halted, hopefully temporarily, by the Covid crisis. Ethiopian, however, used regional aircrafts, whose limited capacity can hardly reduce the costs and customers' prices to a level comparable with Europe or Southeast Asia. At the same time, capacity at airports can be an issue and the presence of a monopolist is typically going in the direction of *keeping* capacity constraints. The real push could come from a strong political commitment towards air transport liberalization, which can give international investors the message that barriers to entry are or will be removed.

4   Concluding Remarks: A Balanced Recipe?

The very different outcomes of the two megaprojects of Maputo province, namely, the Maputo corridor and the Maputo-Catembe Bridge, are symptomatic of their different nature. Both are driven by international investments and interests, but the successful case of the Maputo Corridor shows that a network missing link solution in presence of some conditions (the huge and landlocked industrial production of South Africa) is viable and could benefit in the mid-long term also the local economy. In this case, the road provides good access for previously poorly accessible areas and the port is a benefit for the entire country. On the contrary, the Bridge seems a perfect *white elephant*, if not a tool for land grabbing. In both cases, however, the focus is on industry and extraction and the respondence with local people and businesses is scant.

Developing countries like Mozambique need primarily a reliable road network, reasonably performing and not subject to seasonal disruptions, plus solutions for mass transit in larger cities. Roads are the cheapest infrastructure, but still problematic for a poor country and hardly of interest for foreign investors. It is impressive the extraordinary unbalance between the *entire* national budget for road maintenance and development—300 M$/year[5]—and the size of the planned megaprojects of the RSDIP initiative of the *Corredores de crescimento em Moçambique* (Ross et al. 2014): 5 billion USD just for the transport infrastructure of two (the Nacala and Beira

---

[4] Tourism tends to use non-commercial flights to reach farther destinations, of course at a high price and with tiny volumes.

[5] https://www.economiaemercado.co.mz/artigo/o-mesmo-horizonte-com-novas-estradas-e-des tinos Accessed on 28 May 2021.

ones) of the seven planned corridors.[6] Similarly, air transport, if liberalized, would require limited investments and would be able to guarantee a cheap and effective national-scale mobility in addition to coaches.

In conclusion, a balanced recipe would require rethinking the gigantic plans for infrastructure development toward a system focusing on the local economy, based on trade and services, and not only on extraction and heavy international industry needs. Since money is necessary for that, a way could be to revise the concession system used for the current corridor development, introducing mechanisms of compensations to be spent on the construction and maintenance of branches of the regional road network.

References

Beria P, Grimaldi R, Albalate D, Bel G (2018) Delusions of success: costs and demand of high-speed rail in Italy and Spain. Transp Policy 68:63–79

Castel-Branco CN (2008) Os mega projectos em Moçambique: que contributo para a economia nacional. Paper presented at the Fórum da Sociedade Civil para Indústria Extractiva, Museu de História Natural (Maputo), 27–28 November 2008

CFM (2017, 2019) Informação estatística anual. Annual statistical information. CFM, Maputo

Coutinho PB (2018) Sustainable planning of a new city in Mozambique. Braz J Oper Prod Manag 15(2):270–284

Hakim MM, Merkert R (2016) The causal relationship between air transport and economic growth: empirical evidence from South Asia. J Transp Geogr 56:120–127. https://doi.org/10.1016/j.jtrangeo.2016.09.006

Herzog A (2019) Concept project information document (PID)-Maputo urban transformation project-P171449 (No. PIDC27670). The World Bank, pp 1–0

Laplace I, Latgé-Roucolle C (2016) Deregulation of the ASEAN air transport market: measure of impacts of airport activities on local economies. Transp Res Procedia 14:3721–3730

Laplace I, Lenoir N, Roucolle C (2019) Economic impacts of the ASEAN single aviation market: focus on Cambodia, Laos, Myanmar, The Philippines and Vietnam. Asia Pac Bus Rev 25(5):656–682

Lee JW (2019) ASEAN air transport integration and liberalization: a slow but practical model. In: Hsien PL, Mercurio B (eds) ASEAN law in the new regional economic order. Cambridge University Press, Cambridge

Mussagy IH (2014) Os Mega-Projectos em Moçambique: A conclusão precipitada que pode condenar Moçambique ao fracasso? Revista Electrónica de Investigação e Desenvolvimento 2:1–10

O'Connor K, Fuellhart K, Kim HM (2020) Economic influences on air transport in Vietnam 2006–2019. J Transp Geogr 86(102764):1–8

PwC—PricewaterhouseCoopers (2013) Africa gearing up. Future prospects in Africa for the transportation & logistics industry. PwC, Windhoek

Ross DC (2014) Moçambique em Ascensão: construir um novo dia. International Monetary Fund, Washington DC

Ross DC, Lledo V, Segura-Ubiergo A, Xiao Y, Masha I, Thomas A, Inui K (2014) Capítulo 7. Corredores de desenvolvimento de Moçambique: Plataformas para uma prosperidade partilhada. In: Ross DC (ed) Mozambique rising: building a new tomorrow. International Monetary Fund, Washington DC

---

[6] Total forecasted investments in transport infrastructure are huge. PwC (2013) reports 17 billion USD projects in the pipeline.

Sequeira S, Hartmann O, Kunaka C (2014) Reviving trade routes: evidence from the Maputo Corridor. Discussion papers, 14. SSATP, Washington, USA.

Stein S, Kalina M (2019) Becoming an agricultural growth corridor: African megaprojects at a situated scale. Environ Soc 10(1):83–100

Transnet (2019) Freight rail 2019. Available via Transnet. https://www.transnet.net/InvestorRela tions/AR2019/Freight%20Rail.pdf. Accessed 11 Dec 2021

Society: Maputo, a Case of Social Non-simultaneity? A City Repertoire of Issues

Agostino Petrillo

Abstract The aim of this article is to show the socio-historic complexity of Mozambique's capital city. Maputo presents all the traits of heterogenetic urbanism. It is born as a colonial outpost destined for the exploitation of the country's resources. It long remains a white city, which begins to transform only after the 1975 revolution. The growth in the successive decades is chaotic both from a demographic and spatial point on view, joining destiny with other African cities in not leading in any way to a cohesive and shared idea of a city. Maputo remains the fruit of a clash between the city and the countryside, between the colonial model of rationality and the one of the autochthonous populations. It is a realm of Non-Simultaneity, of an unresolved coexistence of different perceptions of time and views on urban life. In the meantime, the predominance of extraction capitalism multiplies its issues in social, organizational, and infrastructural terms, as informal settlements grow, and the near prospects predict a further rise in the number of inhabitants. This situation implies consequences that are difficult to evaluate under the profile of stability and structuralism of the extremely fragile equilibrium existing now and the possible re-ignition of conflicts that are far from subsided.

1 Introduction

Maputo is a city both extremely distinctive and in other ways very common if considered in the general context of the African continent. In fact, within it one may observe nearly any of the distinctive traits of African and third-world urbanism, traits which in Maputo however are confusingly concentrated, overlapped, and inextricably mixed.

With the urbanization of the planet, in what has been named unambiguously the "urban age", African cities are the ones growing the fastest and in the most scattered manner, disseminating throughout the territory the huge socio-economic inequalities which characterize the countries they are situated in, at the same time showing

A. Petrillo (✉)
Department of Architecture and Urban Studies, Politecnico Di Milano, Milano, Italy
e-mail: agostino.petrillo@polimi.it

237

history's role in defining their borders and image. Physical and geologic/climatic factors, but also human factors such as religion, culture, and politics—not to mention the different colonial heritages and the global economic powers' game—are combined and recombined in ever so different ways defining and redrawing cities' physiognomies.

In this brief excursus, I will try to show what elements make up the uniqueness of Maputo, a uniqueness that, as alluded to earlier, derives from the comprehensive and characteristic combination of all the main processes which shapes Global South cities.

We can easily recognize in Maputo the colonial city, the heterogenetic primate city with its structural dualism, whose birth and development were decided with the interest of the motherland at heart, as shown by the centralization of the population and the economic establishments compared to the overall demography of the country. But we can also identify the scars of dependent urbanism, the legacy of the civil war and the attempt of socialist edification, the informal layout of the city in its wide array of declinations, the fragments of the global city in power (related to the Chinese presence), the existence of gated neighborhoods, and even the recent allusions of gentrification. A unique mix that, as one could easily infer, is also the result of a sum of matters and issues which is not and will not be easy to resolve. To paraphrase a famous quote by Lev Tolstoj's *Anna Karenina* one could say that "all happy cities resemble one another, every unhappy one is unhappy in its own way". It is impossible to find a sole cause to this current *unhappiness* without tangling oneself into a research of the how's and the why's, which risks becoming a *regressus ad infinitum.*

Hence, in the following pages I will try to see how the joining of these typical elements end up characterizing the city making it a unicum, a sort of big repertoire of the urbanism of backwardness, almost a specialized catalogue of the contradictions of African cities, but at the same time making Maputo a universal open-air exhibit of the typologies of inequality and urban deprivation. Within it, in fact, coexist areas urbanized in a "traditional" way (that are "cement" built, equipped with services, running water and electricity, insisting upon an urbanized belt supplied with asphalted and paved roads in which generally reside those with the most economic resources) and a variety of semi-urbanized areas in continuous growth in which the greatest majority of active inhabitants of the center resides. The central vertical nucleus is set against a vast occidental periphery of single floor houses part of which in brick, like in Mafalala, that is situated at a short distance by the tall buildings. Even further there is the endless expanse of reed cities, the sea of the periphery. However, let us try to quickly reconstruct the stages leading to the current situation, considering that reading the whole story of the city through the multi-faceted interplay of discontinuity/continuity is necessary.

2 Maputo's Urban History from an Urban Sociologist Perspective

2.1 The Colonial City

The Hippodamian grid which characterizes numerous Hispanic/Portuguese-founded cities is pleasantly re-adapted in Lourenço Marques, which only after the revolution will be renamed Maputo. The structure of the founding city follows the coastline, drawing up a well-structured colonial city, not without the occasional *coquetterie*, as it is shown by the romantic corners by the seaside or the elegant buildings surrounding the area of the gorgeous station. A commercial outpost town, or *ville comptoir* if one wishes to use the synthetic French expression, which remained a city only for whites, at the margins of which stood the *bairros*, for the natives, following a classic dual model. A city that embodied some sort of exotic dream for the Portuguese, and that up until the 1970s might have been the only city worthy of that name in a country in which at the time, of roughly 10 million inhabitants, only 6% was urbanized. In fact, Lourenço Marques had little more than 200.000 inhabitants by the rise of the revolution (Mendes 1979).

2.2 The Socialist City and the Civil War

The segregated universe of the Portuguese colony ended abruptly in 1975 with the independence and the socialist revolution. The city from white becomes black and many Portuguese colonists leave abandoning their homes, while, on the other hand, a mass migration from the countryside to the city sparks up.

Reaching independence fires up a powerful machine attracting population toward Maputo, which grows exponentially, reaching 750.000 inhabitants just 5 years after the revolution in 1980. The civil war harshly divides the field of battle between the countryside, partially controlled by the guerrilla forces, and the cities controlled by the official government, contributing to providing further drive toward urbanization. Rapidly, a paradoxical saturation of the urban fabric emerges due to the new arrivals, a condition that worries the government at the extent that it even tries to deport the "urban parasites" in specifically built "re-education camps", further proving the growth of a urban marginality difficult to control (Miranda Maloa 2016). During the socialist regime, there were efforts to reorganize the city by introducing elements of city planning, such as the *bairros communais*, conceived according to a similar logic to that of the *sites and services* approach: dividing into lots and assisting self-construction with provided building materials. The objective was to substitute straw buildings lacking running water and plumbing with more structured bricks buildings outfitted with basic infrastructure. A great effort for a state with reduced economic abilities and coping with the effects of a bloody civil war. Isolated programs of "socialist" construction also begin, inspired by soviet architectural models. These will stand as ruins in the modern days.

However, the end of the segregative colonial regime does not put the dichotomy between the built city and the native *bairros* to an end. While the colonial city is no longer exclusively white, it maintains—despite some degradation—a predominancy in respect to its surrounding peripheral fabric (Araujo 2003). Nevertheless, the transition from rural to urban entails a series of other social and cultural transformations.

The cement city disorients newcomers and those assigned with the empty apartments, who find themselves projected in an unusual living situation, and conflict arises between those who arrived first and those who arrived afterward, following a well-known dynamic between the established and the outsiders (Elias and Scotson 1994). It becomes difficult reconciling the needs of the new urban inhabitants and the administration of the capital within a national context in which after the Portuguese fled, 95% of the population is illiterate and the citizens with any type of professional training and education are an extremely slim minority.

As leftovers from the socialist period remain not only the brutalist buildings which interrupt here and there the fabric of pretty colonial villas in some parts of the town center, but also social housing complexes, the attempt to reorganize the urban grid, and some nostalgic name of areas of the town and streets, such as the Bairro Dimitrov and the Avenida Mao Tse Tung.

2.3   The Neoliberal City

The beginning of the 1990s mark another stark break: socialism fades and, with the introduction of a new constitution and of free elections, the democratic era begins: a democracy that quickly embraces the neoliberal ideas of the Washington Consensus and the International Monetary Fund. Naturally, this is a neoliberalism of the South of the World, operating at the margins of big financial capitalism and characterized by the deregulation of public policies and downsizing in public interventions. Hence a new abrupt transition between substantially different eras is configured, a jump bringing new kinds of political and social organizations. Neoliberalism ends up further accentuating the process of urban segregation and extreme differentiation between the outskirts and the town center. While the extension of an irregular unplanned periphery has grown uncontrollably, a number of valorization projects begin in the center, in which new construction spark a sort of gentrification, implying population movement and a re-modeling of the social stratification of the areas involved. However, an even deeper change in the social fabric is proven by the gated neighborhoods, protected by electric fences in which the old and new elites reside.

2.4 The Growth of the Informal City

The so-called "informal city" in continuous growth is, in reality, a number of cities, each different from the other, in which it may even be possible to see the hierarchies based on the levels and the ways fragments of "formal" and "informal" are mixed. The centre's power of attraction continues to spark migration from the rural areas of all Mozambique, producing a series of peculiar forms of spatial organization and articulating different phases of settlement and consolidation. The structures of a self-built city, as it often happens in these situations, are based on paths that lead to different internal stratifications, that are micro-hierarchal and tied to the local topography and the relation with water. The structure, apparently chaotic, hides an overlap of functionality and activity. The great informal area in Maputo evokes AbdouMaliq Simone's most general reflections on the peripheral extensions which surround the cities of the Global South, which arise through complex self-organizational processes through the creation of intricate, but articulate, spaces, in which paths, environments, modes of use, and individual—but also collective—productive activities arise (Simone 2016). Hence, the so-called informal city is a complex territory, developing through invisible structures that hide, behind the apparent chaos, an internal order; a network of social infrastructures which allow its survival and exist unbothered even in the percent of other ways of life (Simone 2004).

2.5 The Rural Within the Urban

In Maputo, a continuous recall between two dimensions exists: the center produces a growing margin, which embraces and engulfs the countryside, sometimes without it entirely ceasing to be such. Bizarre hybridizations of urban and rural pop up even at the margins of gated neighborhoods. City and countryside mix together and overlap configuring languages, ways of life, commercial activities; in the peripheries, the countryside can be smelled all around, it is not merely interstitial scenery or survival. The countryside is within the city due to the persistence of traditional rural activities such as small families' agriculture and cattle breeding and the reappearance of rural housing at the margins of the city. Fragments of old social and settlement forms that become stable components of the informal city. One can simply look at Mafalala, the *bairro de lata*, an expanse of precarious housing covered by metal sheets, hard to read through the lens of a peri-urban area. It's a hyper-dense internal suburb, that surely is not countryside, or rather is no longer countryside, with few and far between trees surviving through the thick self-built fabric; however, it is not a city either, as is shown by its lack of essential amenities such as running water and electricity. This is remedied through alternative systems, typical of rural spaces. The rural is within the urban and interacts with it conditioning mindsets, producing original means of survival, and organizing day-to-day life. Cushion zones, zones of intermediation, and passageways where traditional countryside and village institutions find a renewed and different ability of meaning and action, converge and filter the dissemination of the population from a dispersed countryside toward the extreme condensation of the peripheral *bairros.*

As Henri Lefebvre reminds us, there is never shape without content or content without shape. A reflection which helps elucidating how behind these realities there is always a shape-function (Lefebvre 1970). But what remains of the occidentally shaped city when the centralities are reconfigured and restructured by the encounter between rural and urban? When the "industrial" activities in a broad sense, logistics and agricultural activities, lose a clear localization and intersect one another?

A clash of worlds that in Maputo takes on the traits of a loss of clear bearings and spatial proximity, but also and even more so of the blur of temporal coordinates.

3 A Spatialized Non-Simultaneity

Discussing Maputo's social and urban conditions described above, the category of Non-Simultaneity (*Ungleichzeitigkeit*), as introduced by the philosopher Ernst Bloch in the 1930s, may be of some heuristic usefulness. The idea—in formal and abstract terms—is that of the coexistence, at the same time (simultaneously), of social groups that express or represent different times or that have different dynamics of development. Bloch showed the coexistence in Germany of completely distinct temporalities: the overlap in territories in the proximity of social realities tied to different historical times (Bloch 1985). A category which offers numerous suggestions if used as a lens of political reading of the differences between different spaces making up a city like Maputo, where a sort of still colonial dimension and, in some marginal areas, a village/rural mentality are accosted to the image of a globalized city. The category of Non-Simultaneity represents a way to give an intelligible shape to the coexistence, in the same epoch, of diachronic non-homogenous views of the world; helping in explaining the persistence of other space-times, conceptions, and ways of going far back in time, even to the pre-colonial world. It is a deviation, a break that can be perceived in differentiated rhythms, in the simultaneous presence of extremely different perceptions and uses of the city, in the survival of structures and social relationships, the historicity of which becomes not only an element of individualization of an area but also a possible key to its inequalities. This is a possible way to understand the diversity, the lack of encounters which characterizes a subtly divided city. Of course, other suggestions may also come forward: at the open-air market in front of the station, one may still be able to pick up on the hesitation and the awkward uncertainty of the natives facing the city built in European style, a condition of which Frantz Fanon spoke in "*Les Damnés de la terre*" (1961). While in the self-built areas the rural paths re-emerge, in the open-air markets reigns a kind of suspended time, of atemporality. Once again it was Fanon who noted that the European colonial city is built to block and circumscribe the habitative expression of the local natives. Or to say it in an absolutistic way, its aesthetic law intimidates with its regular blocks and its algid rationalistic abstraction, with its obsessive sanitary hygienism, it closes onto those coming from the countryside in a "cage of guilt" and of inferiorization (Fanon 1961). In Maputo this cage not only is built at a symbolic level but seems operating on a material level, as it acts directly on the bodies' physicality: forcing new routes by establishing no-go areas that are both explicit and implicit, visible and invisible. Forcing new rhythms of life and modes of interaction aside from a general reorganization of spatiality. At the end of the day, Maputo's story is the story of a city born from an external overdetermination, substantially extraneous to the surrounding reality, characterized by an "extractive" function which has influenced its different sceneries. A white city, originally built on racial criteria, which continuously worked not only as means for exclusion, but also as a force shaping relations, simultaneously molding both the urban spaces and the sociocultural dynamics. This is how a bizarre, incoherent society was born, one in which the temporal breaks and spatial models overlap until nearly blending into the moral and social order structures.

4 Conclusions

Mozambique is an economically disadvantaged country that as of now seems destined to remain as such. A capitalism that here has been, since perhaps ever, extractively aimed toward the exploitation of natural resources and an economy dependent on investments and foreign capital, joined with an accumulation model based on unproductive yields and their search, which leads to believe that these economic structures are destined to last in time. Even the ongoing decentralization processes and a new interest of big international investors do not alter the predominant model, whose interests are articulated only locally, following the dynamics that Giovanni Arrighi got a glimpse of in his works on capitalism in hostile environments (Arrighi and Piselli 1987). Maputo is the capital city of this country struggling and toiling with its historical baggage of backwardness, difficult to eradicate. With incumbent and diffused youth poverty in which heterogeneous temporalities and languages coexist, it is a city in constant movement and transformation.

The different periods and phases that Maputo has gone through in its history, from colonialism to socialism to neoliberalism have not solved fundamental issues such as the relationship with the rural dimension and the lack of infrastructure and amenities in the continually expanding peripheries. A place in which the aesthetics of the colonial city and the dawning globalized city go hand in hand with extreme marginalization, but also a place in which life unfolds in all its forms. There is not only suffering and danger, nor is there only the dominion of the economy. Maputo's spaces, in their intricate intertwining of formal and informal, yield nonetheless an intense sense of vitality, of activity, and are enveloped in an intense relational life.

Hence, to be understood, these spaces ask for different means of interpretation of the city, which regardless represent a somewhat appropriate response to a non-functional situation, scarred by an enormous disparity of riches and resources. In which the price of an apartment in the newly constructed luxury complexes by the seaside is between 500 and 1000 times the cost of a small house in Mafalala, only a few kilometers away in a straight line. The town center, the old colonial capital city, continues to be the main area of investment.

In one of his works, about twenty years ago, Paul Jenkins had already clearly outlined the limits of the programs put into place by the government to improve city administration, which further revealed their inadequacy when faced with migrations influx and growth of poverty. They were tragically unrealistic concerning the institutional implications and political balance. According to Jenkins it would be difficult to talk about the redevelopment of Maputo's spaces without considering the issues of social services and infrastructure and dealing with informal residential areas which had all the historical, anthropological, and economic reasons to exist in their current shape (Jenkins 2000). A sensible redevelopment would have to take on these starting conditions and move toward a progressive renovation and integration of this fabric into the more ample urban context.

Given the general background conditions, the informal could do nothing but grow, and far from being able to eliminate it, it is crucial to think of ways to improve it and redevelop it. Creating the basis for new forms of interrelation between the formal and the informal, such as flexible strategies based on the involvement and participation of the inhabitants. Jenkins had understood the situation correctly, and the fact that his suggestions have been just partially and weakly followed produced a situation which now seems difficult to improve upon. The coexistence of different factors represents a giant political and administrative issue: from the lack of public intervention in the informal settlements—mostly left to international cooperation agencies and initiatives—to the failure of the random housing policies put in place so far (Jorge and Tique 2020), till the persistence of processes of social polarization, which continue to see the investments privileging the town center undergoing gentrification, to the expense of a disorganized periphery lacking amenities and infrastructure.

Obviously, the Mozambican government and the city administration have limited resources, as it is often the case in many similar situations in the Global South, and the institutions are largely imperfect, also due to the modest education level of many of the directors and employees. However, a series of interventions starting from a rational use of public resources and more generally rethinking the dynamics of urban development seems to have become not just necessary but inevitable. It seems that this may be the only way to manage a gigantic issue of peripheries still unresolved, a suburban world which grows besieging the city. Up until today the attempts of introducing elements of rational governance have failed leading to a peculiar situation, in which is emerging an indirect way of self-organization in some peri-urban territories. A sort of bottom-up spontaneous planning leading to a condition of *inverse governmentality* intended in Michel Foucault's sense (Nielsen 2011). However, clearly, these are far from sufficient. The timeframe to intervene successfully is not infinite either; a considerable part of Maputo remains without running water or electricity, while prospects warn about a further increase of the population: 11 million expected by 2075 and maybe 21 million by 2100.

References

Araujo M (2003) Os espaços urbanos en Mozambico. GEOUSP Espaço e tempo14:165–182

Arrighi G, Piselli F (1987) Capitalist development in hostile environments: feuds, class struggles, and migrations in a Peripheral Region of Southern Italy. Review (Fernand Braudel Center) 10(4):649–751

Bloch E (1985) Erbschaft dieser Zeit. Werkausgabe Band 4. Erweiterte Ausgabe Suhrkamp, Frankfurt am Main

Elias N, Scotson J (1994) The established and the outsiders: a sociological enquiry into community problems. Sage, London

Fanon F (1961) Les damnés de la terre. Maspero, Paris

Jenkins P (2000) Urban management, urban poverty and urban governance: planning and land management in Maputo Mozambique. Environment & Urbanization 12(1):137–152

Jorge S, Tique J (2020) Fundo para o Fomento à Habitação de quem? Análise do seu impacto a partir do caso da Área Metropolitana de Maputo, Moçambique. Cidades, Comunidades e Territórios 41:209–222. https://doi.org/10.15847/cct.19854

Lefebvre H (1970) Du rural à l'urbain. Anthropos, Paris

Mendes MC (1979) Maputo antes da independencia: geografia de uma cidade colonial. Dissertation. Universidade de Lisboa

Miranda Maloa J (2016) A Urbanização mozambicana. Uma proposta de interpretaçaõ Dissertation. Universidade de Sao Paulo

Nielsen M (2011) Inverse governmentality: the paradoxical production of peri-urban planning in Maputo, Mozambique. Critique of Anthropology 31(4):329–358

Simone AM (2004) People as infrastructure: intersecting fragments in Johannesburg. Public Culture 16(3):407–429

Simone AM (2016) Cities that are just cities. In: Lancione M (ed) Rethinking life at the margins: the assemblage of contexts, subjects, and politics. Routledge, London

Rural/Urban/Metropolitan: Trying to Reduce Inequalities Through Planning

Laura Montedoro

Abstract The regulation of land use is a crucial tool to try to ensure conditions of equity and access to fundamental and non-negotiable rights: living, moving, eating. In the Anglo-Saxon disciplinary tradition of urban planning, this regulation has represented an unavoidable node to try to govern the tumultuous and dynamic processes of metropolisation of the territories, meaning by this expression a field of forces that implies a radical reconfiguration of the relational and settlement framework, characterized by the emergence of new polarities and new gravitation that transform the historical city/countryside relationships. In recent decades, in Sub-Saharan Africa, this process has taken on such strong, accelerated, and uncontrolled forms that it has radically and irreversibly redesigned some centuries-old territorial relations, such as that between urban and rural, and produced new forms of exclusion and inequality in a global world that is increasingly polarized between poor-poor and rich-rich. The contribution, starting from the research experience of "Boa_Ma_Nhã, Maputo!", aims to focus on the main criticalities/inadequacies of planning practices in the field today, reinterpreting the relationship between rural-urban-metropolitan starting from the water-energy-food nexus (WEF nexus), assuming a transcalar and integrated approach as a necessary condition for a strategic vision of sustainable development.

1 Rural/Urban/Metropolitan

The regulation of land use and land infrastructure are crucial tools to try to ensure conditions of equity and access to non-negotiable fundamental rights: living, moving, feeding oneself.[1]

---

[1] Original text in Italian, translated by the author.

L. Montedoro (✉)
Department of Architecture and Urban Studies, Politecnico di Milano, Milan, Italy
e-mail: laura.montedoro@polimi.it

In the Anglo-Saxon disciplinary tradition of urban planning, the so-called rational-comprehensive approach, regulation, and infrastructuring have represented an unavoidable node to try to govern and guide the tumultuous and dynamic processes of metropolisation of the territories, meaning by this expression the outcome of

> the concomitant and interactive action of three factors: the capillary entrance of the market in all the elements that define the framework of life; the technological revolution in transport and telecommunications; the loosening of community relations [...]; a field of forces made up of new potentials and constraints, in continuous evolution, urged/conditioned by public policies (weak) and by private initiative (increasingly strong). This field of forces corresponds to a radical reconfiguration of the relational and settlement framework, characterized by the emergence of new polarities and new gravitations which transform the historical city/countryside relations (Consonni 2020).[2]

In recent decades, this process of metropolisation in Sub-Saharan Africa has taken on forms so rapid, accelerated, and uncontrolled that it has radically and irreversibly redesigned some centuries-old territorial relations, such as that between urban and rural, and to produce new forms of exclusion and inequality in a global world increasingly polarized between poor-poor and rich-rich (Secchi 2013), both in the countryside—where 85% of global poverty is concentrated—and in the big cities, where the last ones arrived (the new citizens) often remain excluded from primary services. Complicating the picture is a senseless and extremely inequitable use of natural resources, foreign to any principle of environmental sustainability, as well as another degree of widespread informality and spontaneity (in local economies, settlements, and mobility) that make the usual planning tools even weaker, if not actually counterproductive (Chiodelli and Tzfadia 2016).

The challenge facing planners and policymakers in this context is therefore very difficult: it is a matter of trying, in fact, to "govern the ungovernable" (Paone et al., 2018). Yet, despite the complexity of the framework just described, there is no lack of some clear and indispensable guidelines regarding the need for urban-rural linkages, widely shared by the international community (UN-Habitat 2019; UN 2015, 2016), which we need to try to pursue urgently.

The research "Boa_Ma_Nhã, Maputo!" was an opportunity to confront these challenges and to develop an agenda for the Province of Maputo, reinterpreting the relationships between rural, urban and metropolitan starting from the nexus between water-energy-food resources (WEF nexus). Through the investigation of this large territory around the Mozambican capital, in constant transformation, some elements of the fragility of the processes of territorial transformation have been identified; they make it more difficult a transcalar and integrated treatment of the context, instead considered a necessary condition for a strategic vision of sustainable development.

---

[2] Original text in Italian, translated by the author.

2   Fragmentation of Knowledge and Planning

Among the main reasons for the lack of effectiveness in the care of urban-rural linkages there is a significant fragmentation of both cognitive actions and the different projects that are deposited on the territory.

In particular, it is possible to recognize some families of actions in poor connection with each other. Maputo and its province are the focus of analyses, policies, and projects that follow different rationales: those of national and local politics; those of international cooperation for development; those of global capital; those of universities; those minimal and individualistic of the populations. Certainly, there is no lack of connections between these different rationales and the effort to seek synergies is certainly high (and the PIMI Project is a promising example), however, there are persistent weaknesses that represent a significant obstacle to a deep understanding of the system of rural-urban-metropolitan relations and the consequent integrated and sustainable planning. Moreover, the analyses on which decision-makers' choices are based are eminently quantitative, which seems paradoxical due to the scarcity and frequent unreliability of data.

2.1   The Weakness of Territorial Governance

Among the critical points noted, certainly the strong sectorization and division of administrative levels produces a weakness in the capacity for territorial government. Existing administrative boundaries and planning tools are inconsistent in respect to the ongoing urbanization patterns, socio-economic dynamics, and correspondent needs. Moreover, governance system appears weak, with a historical lack of interest in urban development at the national government level and limited executive or facilitating capacities at the local level.

The ongoing efforts to re-organize and implement the planning framework at all scales, both in response to the new Territorial Planning Law (2007) or through alternative envisioning processes, are still lacking a trans-scalar and cross-boundary coordination able to offer an integrated territorial decision-making and planning framework. On one hand, the aim and effectiveness of top-down planning processes, relaunched with the approval of the new law, is still extremely narrow as physical planning is seen as a technical problem instead of a potential integrated and synergic process. On the other hand, even the main large-scale strategic vision, related to the commercial relations with South Africa through the Maputo Development Corridor, is not distributing its effects, being just a fast connector to Maputo port, instead of a backbone for a networked trans-scalar system to activate synergies with local actors and communities.

The urgency and complexity of the cross-scalar challenges that Maputo has to face require to support attempts in finding the better solution possible in terms of horizontal and vertical coordination, through strategic and flexible governance frameworks to support alternative planning approaches, in coherence also with the recent recommendations of UN-habitat (UN-Habitat 2019: 8).

> Strengthen governance mechanisms by incorporating urban-rural linkages into multisectoral, multi-level and multi-stakeholder governance. To deliver the SDGs and to address the humanitarian-development-peace nexus, for example, requires policies, strategies and action plans that are: *horizontally integrated* across spatial scales in metropolitan regions, adjacent cities and towns, including rural hinterlands; *sectorally integrated* with the public and private sectors, civil society organizations, research and professional institutions, formal and informal civic associations; and *vertically integrated* across different levels of engagement and official decision making. Enhance and institutionalize synergies from the integration of urban and rural actors and actions across horizontal, sectoral and vertical dimensions.

About the current institutional framework and planning tools, The Government of Mozambique introduced the Territorial Planning Law in 2007, followed by its Regulation in 2008. The Law describes the use of tools and approaches to promote land use plans at National, Provincial, and District level (Monteiro et al. 2017). According to the Law, territorial planning is defined as the process of elaboration of plans that shape spatial forms of relations between humans and their physical and biologic environment, regulating rights and forms of land use and occupation and it's considered one of the main development challenges for Mozambique for the period 2015–2035 (Monteiro et al. 2017; Boa_Ma_Nhã, Maputo! 2020).

Maputo City is a Municipality out of Maputo Province. Maputo Province includes 8 districts, among which Moamba, Boane, and Namaacha. These last two districts include a municipality with the same name, political center of both the district and the municipality. The physical and political relation between districts and municipalities in some cases is particularly challenging.

In fact, municipalities are autonomous entities and there is a legal basis that promotes cooperation between them. In turn the districts depend on the provincial government, the sectoral and central organs of the state. Thus, on the one hand, horizontal relations are established between the municipalities and the state, and on the other hand, vertical relations between the districts, the provincial government, and the central authorities. In this way municipalities and districts do not have a legal and formal framework for integrating their planning tools and visions (Macucule 2015).

In addition, the ongoing and incomplete administrative decentralization have been hindering the horizontal integration and cooperation among the different stakeholders operating in the Maputo metropolitan region and province. Finally, the growing presence and importance of (non-public) actors into the spatial planning arena—i.e., international cooperation agencies and bodies, bit transnational companies—have added further complexity to the already fragmented field of urban governance, often with effect of "down-grading" the role of local authorities in the decision-making process (Macucule 2015).

2.2 The "Realization" of the Metropolis: an Uncertain Urbanity

Despite the dream of a "rational city" (Boyer 1983), the post-colonial transition marked a new cycle of life for Mozambique's capital, full of opportunities but also of pitfalls. Although the exciting post-independence phase had put in place a long-term strategy for the production of new services, the Civil War, the scarcity of resources, and the lack of local expertise gave rise to a process of "urban anarchy". On the one hand, the colonial heritage—in its various forms of the "concrete city", infrastructure, and second homes by the sea or inland—was neglected to the point of complete abandonment; on the other hand, the "city of caniço", or so-called informal settlements, continued to grow by leaps and bounds, without ensuring access to primary services or an efficient public transport system (Cfr. Chap. 1; Jenkins 2003; Jenkins 2021).

Even though the policy of "villagization" (Lorgen 1999) had invested in the model of an agricultural-based country (Borges Coelho 1998), the city remained an aspiration for much of the population. The consistency and speed of the phenomena of urbanization have thus produced the Maputo that we see today: a well-designed but largely decaying hard core, coinciding with the colonial city, and an extension as far as the eye can see of a low-density and predominantly informal periphery, extending well beyond the municipal administrative limits (see Chap. 5), to form an impressive conurbation with municipalities (see Matola and Boane) and districts (Marracuene) nearby.

A phenomenon similar to the *sprawl* of some European cities—of which it has all the criticalities relating to land consumption and settlement dispersion: removal of productive and natural land; erosion and loss of landscape quality; indifferent disposition to the different characteristics of vulnerability of the territory, with consequent increase in damage suffered and caused; greater burdens in the distribution of services; an increase in mobility based exclusively on road transport of goods and people and the impossibility of providing an adequate collective transport service (Di Gennaro and Innamorato 2005; Indovina 2009)—but in a framework made more complex and problematic by the particular regime of land use, the scarcity of some primary services, the absence of public transport, the irrational use of resources. In rethinking rural-urban/metropolitan relations, it is therefore a priority to imagine alternative settlement models, both to improve the quality of the existing city and to govern the growth of the city, preserving precious and endangered environmental systems.

Observing the so-called "Greater Maputo"—a still unofficial territorial entity (Cfr. Chap. 5)—we also wonder about the forms of *urbanity* in this context: of the most widely shared definition, it certainly maintains the intensity of the relations between inhabitants, but lacks many other characteristics, such as the functional and social mix, the variety and liveability of public spaces, the widespread supply of services, the high level of accessibility, and the diffuse quality of places. On the other hand, we find some peculiar features that we find difficult to classify in the category *urbanity*:

the prevalence of unpaved streets and public spaces; the ever-present urban home vegetable gardens, created within the family housing unit; the widespread breeding of farmyard animals, always within the house; the predominance of informal street commerce; it is, in other words, a *ruralized* urbanity. A metropolitan context that has brought within itself an atomized, molecular countryside, often indispensable for families who thus ensure a certain source of self-sustenance. This *uncertain urbanity* is another important factor to take into account when rethinking urban-rural linkages.

2.3 The Impact of Global Capital on Land Use Infrastructure

Finally, for a planning action aimed at rebalancing the relationship between town and country, the effect of exogenous forces outside the logic of local development cannot be ignored. These forces may be public, such as the government of a foreign country, or private, such as the large multinational companies, and may act directly or indirectly; in both cases, they are expressions of the global economy, typical of mature capitalism, in action everywhere on the planet, often with a certain opacity of the processes.

In the Mozambican context, it is possible to recognize at least three different forms of influence/intervention that have direct effects on the transformation of the territory. The first and perhaps best known is that of actions aimed at the exploitation of natural resources, especially mining; in this case, the effects can be direct, with the construction of ad hoc infrastructures, functional to the enterprise (new roads, gated communities for foreign personnel, platforms, etc.) or indirect, i.e., subject to negotiation with local institutions. The second is the form of land-grabbing: the massive and extensive purchase of land for food production destined for the investiture country (Cotula et al. 2009; cfr. Chap. 8). The third is the construction of mobility infrastructures designed to guide development processes, often on transnational axes; in Maputo Province this is the case of the recent bridge over the bay, linking the city to Katembe, built by Chinese cooperation, and the now consolidated Maputo-Johannesburg corridor with the construction of the N4 motorway.

The impact of these projects is considerable, not only in terms of their effects on the territories through which they pass, but above all in terms of the radical changes in the systems of relations on a large scale. On the other hand, the infrastructuring of the territory has always been a very powerful tool for guiding development and transformations, producing new proximities but also new distances; structuring, in other words, new hierarchies which, unless very careful and balanced actions are taken, generate exclusions, inequalities, and new poverty.

It is therefore unimaginable to think in terms of rural/urban/metropolitan without imagining a policy to contain the forces of global capital that are redesigning entire regions for purposes that are certainly not aimed at territorial rebalancing, not least to ensure that Mozambique has supplies and its own food reserves.

Although it is beyond the scope of this paper, it should also be noted that all these actions involve more or less intensive processes of resettlements, with the uprooting of local communities, often relocated a long way from their original homes (Beier et al. 2021).

3 Concluding Remarks and Open Prospects

In sub-Saharan Africa, rural areas are becoming increasingly impoverished and are tending to be depopulated in favor of the large poles of attraction of the metropolises, which seem to offer better living conditions in terms of opportunities (prospects for economic and social upgrading) and better access to primary goods and services.

In Maputo Province, a phenomenon of metropolisation is very visible: the capital and its surroundings continue to grow, without due attention either to the sustainability of the urban growth model or to the natural and rural environment. On the other hand, the countryside is not equipped to guarantee genuinely profitable productivity and competes with the city for certain primary resources (water above all; Cfr. Chap. 8). Between them, there is a constellation of small and medium-sized towns that do not actively trade with the flow of goods and people, from which they are largely excluded, but which have a fair amount of public services, sometimes underused.

It is therefore necessary to develop a comprehensive, integrated and multidisciplinary vision so that

> Urban and rural areas should not be treated as separate entities when development plans, policies and strategies are made. Rather, the aim is to harness the potential that their combined synergy generates, so that everyone benefits from the circular flow along the urban-rural continuum (UN-Habitat 2019).

In order to achieve this ambitious objective, in addition to consolidating the cognitive analysis of the territory through the collection of more complete and reliable data (quantitative analysis), it is essential to also put in place a capacity for qualitative analysis, through intense observation and survey work in the field, as the Boa_Ma_Nhã, Maputo! has tried to experiment.

References

Beier R, Spire A, Bridonneau M (2021) Urban resettlements in the global south: lived experiences of housing and infrastructure between displacement and relocation. Routledge, London
Boa_Ma_Nhã, Maputo! (2020) Planning tools report. Politecnico di Milano, Milan
Boyer MC (1983) Dreaming the rational city. The MIT Press, Cambridge, MA
Borges Coelho JP (1998) State resettlement policies in Post-Colonial Rural Mozambique. J South Afr Stud 24(1). Special Issue on Mozambique: 61–91
Consonni G (2020) Città e metropoli: lo scenario ridisegnato dalla pandemia, «Arcipelago Milano»
Cotula L, Vermeulen S, Leonard R, Keeley J (2009) Land grab or development opportunity? Agricultural investment and international land deals in Africa. FAO, IIED, IFAD, London/Rome
Chiodelli F, Tzfadia E (2016) The multifaceted relation between formal institutions and the production of informal urban spaces. An editorial introduction. Geogr Res Forum 36:1–14

Di Gennaro A, Innamorato FP (2005) La grande trasformazione. Il territorio rurale della Campania 1960/2000, Clean Edizioni, Napoli

Indovina F (2009) Dalla città diffusa all'arcipelago metropolitan. Franco Angeli, Milano

Jenkins P (2003) In search of the urban–rural frontline in postwar Mozambique and Angola. Environ Urban 15(1):121–134. https://doi.org/10.1177/095624780301500115

Jenkins P (2021) Colonial and post-colonial continuities and discontinuities in urban infrastructure in Africa: a case study in Maputo. In: Vaz Milheiro A (ed) Coast to coast—late Portuguese infrastructural development in continental Africa (Researchers' book). AMDJAC, Porto

Lorgen C (1999) The experience of Villagization: lessons from Ethiopia, Mozambique and Tanzania. OXFAM, London

Macucule D (2015) Processo-Forma Urbana: Restruturação urbana e governança em Maputo. Dissertation. New University of Lisbon

Monteiro J, Inguane A, Oliveira E, Joaquim S, Matlava L (2017) Territorial planning at community level in mozambique: opportunities and challeges in a context of community land delimitation. 2017 World Bank conference on land and poverty. The World Bank, Washington DC

Paone S, Petrillo A, Chiodelli F (2018) Governare l'ingovernabile. Politiche degli slum nel XXI secolo. Edizioni ETS, Pisa

Secchi B (2013) La città dei ricchi e la città dei poveri. Laterza, Roma-Bari

UN - United Nations (2015) Transforming our world: the 2030 agenda for sustainable development. United Nations, New York

UN - United Nations (2016) New urban agenda. United Nations, New York

UN-Habitat (2019) Urban-rural linkages: guiding principles. Framework for action to advance integrated territorial development. UN-Habitat, Nairobi

Afterword. Learning from Research by Design Approaches for a WEF-Sensitive Planning Culture

Gabriele Pasqui

This book represents a stage in the long-established cooperation between the Politecnico di Milano—and, in particular, the Department of Architecture and Urban Studies (DAStU)—and different Mozambican stakeholders.

Many projects and collaborations promoted in the last 5 years have created the conditions for a stable collaboration between DAStU and the Faculty of Architecture and Physical Planning (Faculdade de Arquitectura e Planeamento Físico—FAPF) of the Eduardo Mondlane University. This fertile interaction resulted in our collaboration with the *Integrated and Multisectoral Research Program: Study for the Promotion of Integrated Territorial Development in the Region of Boane, Moamba and Namaacha* (*Programa de Investigação Multissectorial Integrada: Estudo para a Promoção do Desenvolvimento Territorial Integrado da Região de Boane, Moamba and Namaacha*—PIMI), a program financed by the Italian cooperation and especially by the Italian Agency for Development Cooperation (AICS) in Maputo. For all these reasons, the research presented in this book, among which the Polisocial Award "Boa_Ma_Nhã-Maputo!" represents a meaningful example, is both a resume of a work done in many years and a new starting point.

These characteristics of the research narrated in this book help us understand why social engagement by European universities in the African context is both necessary and difficult. I was able to reflect more directly on these difficulties thanks to the visits to Maputo I have been involved in during the past 4 years.

The first time I visited Maputo, in February 2018, I understood something about the city, about Mozambique more in general, and about sub-Saharan Africa at large. Not much, actually, because to understand places and territories it takes patience, time, and a capability to listen and watch that requires training and determination.

G. Pasqui (✉)
Department of Architecture and Urban Studies, Politecnico di Milano, Milano, Italy
e-mail: gabriele.pasqui@polimi.it

© The Author(s), under exclusive license to Springer Nature Switzerland AG 2022
L. Montedoro et al. (eds.), *Territorial Development and Water-Energy-Food Nexus in the Global South*, Research for Development,
https://doi.org/10.1007/978-3-030-96538-9_19

Among the things I understood on that first trip, also confirmed by the other two periods I spent in Mozambique between 2018 and 2019, is that by encountering worlds, places, and spaces that are very different from ours, we are called to respond to a challenge. The challenge concerns the way in which we expose ourselves, with our history, language, culture, and tradition to contexts and people that, at a first glance, seem very different to us and question our forms of rationality.

Time, space, rhythms of the city seem very far from those we inhabit in European cities, and the temptation to impose rationalization logics is very strong.

To respond to the challenge, however, we must never forget that Maputo and its region, and Mozambique more in general, are also the result of processes of anthropization and urbanization that are profoundly marked by the culture and knowledge of the Western culture, in this case by Portuguese colonialism, but also by the English one. How can we forget the rigorous grid of the Portuguese city? On the other side, we cannot ignore a sense of *otherness*, which is anthropological, social, and economic, but also material: the red soil; the unknown and gigantic plants that grow even in the city; and forms of human and non-human life that sometimes leave us amazed and that coexist with the artifacts, products, and objects of globalization.

The link between human beings and natural conditions, which is at the center of the reflections proposed by this volume, assumes shapes and dimensions in Mozambique that are different from those we are used to, even in areas with the strongest urbanization such as that of Maputo.

Without forgetting the perception of injustice and poverty, which I experienced personally during my first visit to the informal settlements of the city. Poverty and injustice should not leave us indifferent and should oblige us to a responsibility that is not (only) personal, but also collective, and must be linked to our skills and our work as planners and researchers.

Never as in recent months, in the midst of the global pandemic crisis in which the news on the spread of the virus in Africa remains uncertain, we understand how the ecological issues addressed in this book are strongly political. The Water-Energy-Food Nexus is an opportunity to rethink the development model of the urban region of Maputo, and to promote an ecological transition that can affect radically the dynamics and the power relations governing the processes of urbanization. The issues addressed in this book are, therefore, closely intertwined with the need to deeply rethink the urban and socio-economic development strategies in the context of Maputo and throughout Mozambique, promoting socio-spatial equality as well as sustainable development trajectories.

As it is extensively explained in this beautiful book, also the experience of the project "Boa_Ma_Nhã, Maputo!" starts in the context of the cooperation framework of the PIMI project, although it also presents an autonomous research path, enriched by the interdepartmental cooperation characterizing the projects financed by Politecnico through the Polisocial Awards program.

"Boa_Ma_Nhã, Maputo!" has been carried with passion, competences, managerial skills, and it represents a significant example of academic social engagement in the context of development cooperation for several reasons.

First of all, the research tries to make knowledge, skills, and passion available to different actors and stakeholders in a very complex but exceptionally interesting context such as that of Mozambique. It does so by assuming a partially unprecedented design perspective in the Mozambican context, building the conditions for a fruitful cultural exchange.

Secondly, this research has been carried out with full awareness of the risks involved in knowledge transfer in a context such as the sub-Saharan African one. Precisely for this reason, the research was based on a very thorough work of direct knowledge, exploration of places, and dialogue with local actors. Days and weeks spent in Maputo and in Boane, Moamba e Namaacha districts have been a crucial part of the research tools and a condition of realism of the conclusions reached during the research.

Thirdly, it is a research initiative that aims to open new ways of co-design and collaboration, through the consolidation of a promising policy network open to institutional actors and NGOs. "Boa_Ma_Nhã, Maputo!" is an example of a research project in the field of development cooperation deeply rooted in its context and based on a strong interaction with the relevant local actors.

Finally, the transdisciplinary dimension of the work carried out in Maputo is very well highlighted in the volume. The cultures of urban planning and urban design, environmental and energy engineering, and ICTs converge not only to provide an original interpretation of the metropolization processes of the Maputo area, but also to highlight possible strategies and guidelines for the main Mozambican urban region.

The book edited by Laura Montedoro, Alice Buoli, and Alessandro Frigerio sets out clearly the intentions and objectives, process, and results of the research path, which I consider very significant not only for the operational contribution offered to the PIMI program and more generally to the action of territorial planning in the Maputo area, but also for what "Boa_Ma_Nhã, Maputo!" teaches us, planners and urban designers trained and active mainly in western academic contexts.

I would like, therefore, to try identifying seven lessons that, in my opinion, the project and this book give us, for further reflection.

1. To work in contexts such as that of the Maputo Region we need time. Time to observe, map, and analyze the territory and its characters. As very well shown in the book, we need different kinds of data: physical and socio-economic. And we must try to construct representations that allow us to read and interpret them together. This is not an easy task: gathering information is very hard work, and at the same time necessary. Never as in this case, it is essential to start from quantitative evidence, from an approach that is as analytical as possible. Not because design choices can mechanically derive from information. Rather, because the production of information is not only an output of the research process, but also a crucial tool for building an active dialogue with our interlocutors, a means of creating a common ground among institutions, academics, social actors, and citizens. For these reasons, in this research and in this book, data should be considered also as policy tools and mechanisms.

2.   We need technical knowledge. Therefore, the collaboration between architects, urban planners, energy and hydraulic engineering experts, and scholars in ICT modeling takes on such an important role. The transdisciplinary dimension of "Boa_Ma_Nhã, Maputo!" project constitutes not only an opportunity to build a robust framework to address the decisive issue of the Water-Energy-Food (WEF) Nexus, but also a legitimizing device for planning choices, in a context in which the acquisition of technical skills represents a decisive step to give strength to any prospect of change. In this perspective, the use of technical knowledge, the connection with the *usable knowledge* produced in the research process, and the mechanisms of its production are far more important than the specific analyses.

3.   We should use and practice a multiplicity of views. We must be able to observe from far above, grasping the relationships on a transnational scale (the connections between the development of the Maputo area and the infrastructure of the Maputo-Johannesburg corridor is extensively treated in the volume), but we must also be able to recognize the specificities and variety at the scale of the urban region: Boane, Moamba, and Namaacha have very different trajectories and characteristics, which this research has recognized and enhanced. This transcalarity of the research perspective also requires different tools, which must be able to communicate effectively with each other.

4.   We must be able to deliver to our interlocutors not top-down plans (often very defined, rarely effective), but operational tools. In fact, the authors defined their work as a "research by design project". The scenarios, the recommendations, the guidelines, and the illustrative actions constitute a form of outcome and communication of the results of the design research that avoids the rigidity of plans and programs that are often disregarded. These different outcomes try to build the conditions for a change that is, first of all, made of innovations in representations, imaginaries, technical cultures, and political strategies. From this point of view, "Boa_Ma_Nhã, Maputo!" has been an attempt to experiment a new planning style in a context characterized by radical environmental, institutional, and political uncertainty, where top-down approaches have demonstrated to be often completely ineffective.

5.   We must plan and test our hypotheses and our readings by concretely experimenting, in specific places, what we have learned, also assuming a perspective capable of offering concrete examples of how a territory can effectively change. For this reason, images are crucial, and data are not enough: we need to show the consequences of our strategies and actions on everyday life. It is not always easy to take daily life as a privileged point of view. In fact, this implies a culture of planning that takes as a perspective that of effects and not of (good) intentions. A culture of this type also assumes that our activity must always be commensurate with the populations who inhabit, live, and move around the territory, for example, in the Maputo Region.

6.   We must be aware of the need to address governance issues. Without actors, there are no public policies. For this reason, we must map not only places and resources, but also institutional and non-institutional actors, interests (even in conflict with each other), and potential alliances between local and non-local

actors. That is, we must be aware that the work produced is delivered to a field of multi-actor social interaction and can be used in many ways. For this reason, we must not imagine that what we have produced can be immediately used: it will produce effects, if it does, only over time and based on processes that are only in our hands to a small extent.

7. Finally, we need to know very well how difficult this task is. "Boa_Ma_Nhã, Maputo" and PIMI have been proposed, above all, as an empowerment device for public administrations, as an opportunity to promote public discussion, as a terrain for potential enrichment of networks between actors. In this book too, awareness of the difficulties emerges clearly, showing the importance of an attitude that knows how to be both modest and visionary, realistic, and open to possibilities.

When, between August and September 2020, in full pandemic and therefore using a distance learning setting, I carried out my Strategic Spatial Planning module as part of the PIMI program, working with a class of about 10 Ph.D. students and 20 students of Masters from Mondlane University, I have found that many of the results of the work done with "Boa_Ma_Nhã, Maputo!" and the other research contributions presented in this volume can become a common basis for reflection for a new generation of Mozambican planners and urban designers. On the other hand, in contact with these young people, I understood how much we can learn from their experience, from their knowledge, and from their closer look.

Milton Keynes UK
Ingram Content Group UK Ltd.
UKHW020112280923
429512UK00001B/17